Hit Refresh

On the morning of February 4, 2014, I was introduced to employees as Microsoft's third CEO alongside Bill Gates and Steve Ballmer, the only CEOs in Microsoft's forty-year history.

Hit
Refresh

The Quest to Rediscover
Microsoft's Soul and
Imagine a Better
Future for Everyone

Satya Nadella

with Greg Shaw and Jill Tracie Nichols

**WILLIAM
COLLINS**

William Collins
An imprint of HarperCollins*Publishers*
1 London Bridge Street
London SE1 9GF

WilliamCollinsBooks.com

First published in Great Britain in 2017 by William Collins

1

A catalogue record for this book is available from the British Library

ISBN 978-0-00-824765-2 (hardback)
ISBN 978-0-00-824766-9 (trade paperback)

Designed by Bonni Leon-Berman

Printed and bound in Great Britain by CPI Group (UK) Ltd, Croydon CR0 4YY

MIX
Paper from
responsible sources
FSC™ C007454

This book is produced from independently certified FSC™ paper
to ensure responsible forest management.

For more information visit: www.harpercollins.co.uk/green

To the two families that have shaped my life:

Anu, our parents and our children;

and my Microsoft family

Contents

Contents

Foreword

BY BILL GATES

I've known Satya Nadella for more than twenty years. I got to know him in the mid-nineties, when I was CEO of Microsoft and he was working on our server software, which was just taking off at the time. We took a long-term approach to building the business, which had two benefits: It gave the company another growth engine, and it fostered many of the new leaders who run Microsoft today, including Satya.

Later I worked really intensely with him when he moved over to run our efforts to build a world-class search engine. We had fallen behind Google, and our original search team had moved on. Satya was part of the group that came in to turn things around. He was humble, forward-looking, and pragmatic. He raised smart questions about our strategy. And he worked well with the hard-core engineers.

So it was no surprise to me that once Satya became Microsoft's CEO, he immediately put his mark on the company. As the title of this book implies, he didn't completely break with the past—when you hit refresh on your browser, some of what's on the page stays the same. But under Satya's leadership, Microsoft has been able to transition away from a purely Windows-centric approach. He led the adoption of a bold new mission for the company. He is part of a constant conversation, reaching out to customers, top researchers, and executives. And, most crucially, he is making big bets on a few key technologies, like artificial intelligence and cloud computing, where Microsoft will differentiate itself.

It is a smart approach not just for Microsoft, but for any company that wants to succeed in the digital age. The computing industry has never been more complex. Today lots of big companies besides Microsoft are doing innovative work—Google, Apple, Facebook, Amazon, and others. There are cutting-edge users all around the world, not just in the United States. The PC is no longer the only computing device, or even the main one, that most users interact with.

Despite all this rapid change in the computing industry, we are still at the beginning of the digital revolution. Take artificial intelligence (AI) as an example. Think of all the time we spend manually organizing and performing mundane activities, from scheduling meetings to paying the bills. In the future, an AI agent will know that you are at work and have ten minutes free, and then help you accomplish something that is high on your to-do list. AI is on the verge of making our lives more productive and creative.

Innovation will improve many other areas of life too. It's the biggest piece of my work with the Gates Foundation, which is focused on reducing the world's worst inequities. Digital tracking tools and genetic sequencing are helping us get achingly close to eradicating polio, which would be just the second human disease ever wiped out. In Kenya, Tanzania, and other countries, digital money is letting low-income users save, borrow, and transfer funds like never before. In classrooms across the United States, personalized-learning software allows students to move at their own pace and zero in on the skills they most need to improve.

Of course, with every new technology, there are challenges. How do we help people whose jobs are replaced by AI agents and robots? Will users trust their AI agent with all their information? If an agent could advise you on your work style, would you want it to?

That is what makes books like *Hit Refresh* so valuable. Satya has charted a course for making the most of the opportunities created by technology while also facing up to the hard questions. And he offers his own fascinating personal story, more literary quotations than you might expect, and even a few lessons from his beloved game of cricket.

We should all be optimistic about what's to come. The world is getting better, and progress is coming faster than ever. This book is a thoughtful guide to an exciting, challenging future.

Hit Refresh

CHAPTER 1

From Hyderabad to Redmond

How Karl Marx, a Sanskrit Scholar, and a Cricket Hero Shaped My Boyhood

I joined Microsoft in 1992 because I wanted to work for a company filled with people who believed they were on a mission to change the world. That was twenty-five years ago, and I've never regretted it. Microsoft authored the PC Revolution, and our success—rivaled perhaps only by IBM in a previous generation—is legendary. But after years of outdistancing all of our competitors, something was changing—and not for the better. Innovation was being replaced by bureaucracy. Teamwork was being replaced by internal politics. We were falling behind.

In the midst of these troubled times, a cartoonist drew the

Microsoft organization chart as warring gangs, each pointing a gun at another. The humorist's message was impossible to ignore. As a twenty-four-year veteran of Microsoft, a consummate insider, the caricature really bothered me. But what upset me more was that our own people just accepted it. Sure, I had experienced some of that disharmony in my various roles. But I never saw it as insolvable. So when I was named Microsoft's third CEO in February 2014, I told employees that renewing our company's culture would be my highest priority. I told them I was committed to ruthlessly removing barriers to innovation so we could get back to what we all joined the company to do—to make a difference in the world. Microsoft has always been at its best when it connects personal passion to a broader purpose: Windows, Office, Xbox, Surface, our servers, and the Microsoft Cloud—all of these products have become digital platforms upon which individuals and organizations can build their own dreams. These were lofty achievements, and I knew that we were capable of still more, and that employees were hungry to do more. Those were the instincts and the values I wanted Microsoft's culture to embrace.

Not long into my tenure as CEO, I decided to experiment with one of the most important meetings I lead. Each week my senior leadership team (SLT) meets to review, brainstorm, and wrestle with big opportunities and difficult decisions. The SLT is made up of some very talented people—engineers, researchers, managers, and marketers. It's a diverse group of men and women from a variety of backgrounds who have come to Microsoft because they love technology and they believe their work can make a difference.

At the time, it included people like Peggy Johnson, a former engineer in GE's military electronics division and Qualcomm executive, who now heads business development. Kathleen Hogan, a former Oracle applications developer who now leads human resources and is my partner in transforming our culture. Kurt Delbene, a veteran Microsoft leader who left the company to help fix Healthcare.gov during the Obama administration and returned to lead strategy. Qi Lu, who spent ten years at Yahoo and ran our applications and services business—he held twenty U.S. patents. Our CFO, Amy Hood, was an investment banker at Goldman Sachs. Brad Smith, president of the company and chief legal officer, was a partner at Covington and Burling—remembered to this day as the *first* attorney in the nearly century-old firm to insist as a condition of his employment in 1986 that he have a PC on his desk. Scott Guthrie, who took over from me as leader of our cloud and enterprise business, joined Microsoft right out of Duke University. Coincidentally, Terry Myerson, our Windows and Devices chief, also graduated from Duke before he founded Intersé—one of the first Web software companies. Chris Capossela, our chief marketing officer, who grew up in a family-run Italian restaurant in the North End of Boston, and joined Microsoft right out of Harvard College the year before I joined. Kevin Turner, a former Wal-Mart executive, who was chief operating officer and led worldwide sales. Harry Shum, who leads Microsoft's celebrated Artificial Intelligence and Research Group operation, received his PhD in robotics from Carnegie Mellon and is one of the world's authorities on computer vision and graphics.

I had been a member of the SLT myself when Steve Ballmer

was CEO, and, while I admired every member of our team, I felt that we needed to deepen our understanding of one another—to delve into what really makes each of us tick—and to connect our personal philosophies to our jobs as leaders of the company. I knew that if we dropped those proverbial guns and channeled that collective IQ and energy into a refreshed mission, we could get back to the dream that first inspired Bill and Paul—democratizing leading-edge computer technology.

Just before I was named CEO, our home football team—the Seattle Seahawks—had just won the Super Bowl, and many of us found inspiration in their story. The Seahawks coach, Pete Carroll, had caught my attention with the hiring of psychologist Michael Gervais, who specializes in mindfulness training to achieve high-level performance. It may sound like Kumbaya, but it's far from it. Dr. Gervais worked with the Seahawks to fully engage the minds of players and coaches to achieve excellence on the field and off. Like athletes, we all navigate our own high-stakes environments, and I thought our team could learn something from Dr. Gervais's approach.

Early one Friday morning the SLT assembled. Only this time it was not in our staid, executive boardroom. Instead we gathered in a more relaxed space on the far-side of campus, one frequented by software and game developers. It was open, airy, and unpretentious. Gone were the usual tables and chairs. There was no space to set up computers to monitor never-ending emails and newsfeeds. Our phones were put away—jammed into pants pockets, bags, and backpacks. Instead we sat on comfortable couches in a large circle. There was no place to hide. I opened the meeting by asking everyone to suspend judgment and try to stay in the moment. I was hopeful, but I was also somewhat anxious.

For the first exercise Dr. Gervais asked us if we were interested in having an extraordinary individual experience. We all nodded yes. Then he moved on and asked for a volunteer to stand up. Only no one did, and it was very quiet and very awkward for a moment. Then our CFO, Amy Hood, jumped up to volunteer and was subsequently challenged to recite the alphabet, interspersing every letter with a number—A1B2C3 and so forth. But Dr. Gervais was curious: Why wouldn't everyone jump up? Wasn't this a high-performing group? Didn't everyone just say they wanted to do something extraordinary? With no phones or PCs to look at, we looked down at our shoes or shot a nervous smile to colleagues. The answers were hard to pull out, even though they were just beneath the surface. Fear: of being ridiculed; of failing; of not looking like the smartest person in the room. And arrogance: I am too important for these games. "What a stupid question," we had grown used to hearing.

But Dr. Gervais was encouraging. People began to breathe more easily and to laugh a little. Outside, the grayness of the morning brightened beneath the summer sun and one by one we all spoke.

We shared our personal passions and philosophies. We were asked to reflect on who we are, both in our home lives and at work. How do we connect our work persona with our life persona? People talked about spirituality, their Catholic roots, their study of Confucian teachings, they shared their struggles as parents and their unending dedication to making products that people love to use for work and entertainment. As I listened, I realized that in all of my years at Microsoft this was the first time I'd heard my colleagues talk about themselves, not exclusively about business matters. Looking around the room, I even saw a few teary eyes.

When it came my turn, I drew on a deep well of emotion and began to speak. I had been thinking about my life—my parents, my wife and children, my work. It had been a long journey to this point. My mind went back to earlier days: as a child in India, as a young man immigrating to this country, as a husband and the father of a child with special needs, as an engineer designing technologies that reach billions of people worldwide, and, yes, even as an obsessed cricket fan who long ago dreamed of being a professional player. All these parts of me came together in this new role, a role that would call upon all of my passions, skills, and values—just as our challenges would call upon everyone else in the room that day and everyone else who worked at Microsoft.

I told them that we spend far too much time at work for it not to have deep meaning. If we can connect what we stand for as individuals with what this company is capable of, there is very little we can't accomplish. For as long as I can remember, I've always had a hunger to learn—whether it be from a line of poetry, from a conversation with a friend, or from a lesson with a teacher. My personal philosophy and my passion, developed over time and through exposure to many different experiences, is to connect new ideas with a growing sense of empathy for other people. Ideas excite me. Empathy grounds and centers me.

Ironically, it was a lack of empathy that nearly cost me the chance to join Microsoft as a young man some twenty years before. Looking back to my own interview process decades ago, I remember that after a full day of interviews with various engineering leaders who tested my fortitude and my intellectual chops, I met Richard Tait—an up-and-coming manager who went on to found Cranium games. Richard didn't give me an engineering problem to solve on the whiteboard or a complex coding scenario to talk through. He didn't grill me on my prior experiences or educational pedigree. He had one simple question.

"Imagine you see a baby laying in the street, and the baby is crying. What do you do?" he asked.

"You call 911," I replied without much forethought.

Richard walked me out of his office, put his arm around me, and said, "You need some empathy, man. If a baby is laying on a street crying, pick up the baby."

Somehow, I got the job anyway, but Richard's words have remained with me to this day. Little did I know then that I would soon learn empathy in a deeply personal way.

It was just a few short years later that our first child, Zain, was born. My wife, Anu, and I are our parents' only children, and so you can imagine there had been much anticipation of Zain's birth. With help from her mom, Anu had been busily equipping the house for a new happy and healthy baby. Our preoccupations were more centered around how quickly Anu might return to her burgeoning career as an architect from maternity leave. Like any parent, we thought about how our weekends and vacations would change when we turned parents.

One night, during the thirty-sixth week of her pregnancy,

Anu noticed that the baby was not moving as much as she was accustomed to. So we went to the emergency room of a local hospital in Bellevue. We thought it would be just a routine checkup, little more than new parent anxiety. In fact, I distinctly remember feeling annoyed by the wait times we experienced in the emergency room. But upon examination, the doctors were alarmed enough to order an emergency cesarean section. Zain was born at 11:29 p.m. on August 13, 1996, all of three pounds. He did not cry.

Zain was transported from the hospital in Bellevue across Lake Washington to Seattle Children's Hospital with its state-of-the-art Neonatal Intensive Care Unit. Anu began her recovery from the difficult birth. I spent the night with her in the hospital and immediately went to see Zain the next morning. Little did I know then how profoundly our lives would change. Over the course of the next couple of years we learned more about the damage caused by utero asphyxiation, and how Zain would require a wheelchair and be reliant on us because of severe cerebral palsy. I was devastated. But mostly I was sad for how things turned out for me and Anu. Thankfully, Anu helped me to understand that it was not about what happened to me. It was about deeply understanding what had happened to Zain, and developing empathy for his pain and his circumstances while accepting our responsibility as his parents.

Being a husband and a father has taken me on an emotional journey. It has helped me develop a deeper understanding of people of all abilities and of what love and human ingenuity can accomplish. As part of this journey I also discovered the teachings of India's most famous son—Gautama Buddha. I am not particularly religious, but I was searching and I was curious why so few people in India have been followers of Buddha despite his origins. I discovered Buddha did not set out to found a world religion. He set out to understand why one suffers. I learned that only through living life's ups and downs can you develop empathy; that in order not to suffer, or at least not to suffer so much, one must become comfortable with impermanence. I distinctly remember how much the "permanence" of Zain's condition bothered me in the early years of his life. However, things are always changing. If you could understand impermanence deeply, you would develop more equanimity. You would not get too excited about either the ups or downs of life. And only then would you be ready to develop that deeper sense of empathy and compassion for everything around you. The computer scientist in me loved this compact instruction set for life.

Don't get me wrong. I am anything but perfect and for sure not on the verge of achieving enlightenment or nirvana. It's just that life's experience has helped me build a growing sense of empathy for an ever-widening circle of people. I have empathy for people with disabilities. I have empathy for people trying to make a living from the inner cities and the Rust Belt to the developing countries of Asia, Africa, and Latin America. I have empathy for small business owners working to succeed. I have empathy for any person targeted with violence and hate because of the color of his or her skin, what they believe, or who they love. My passion is to put empathy at the center of everything I pursue—from the products we launch, to the new markets we enter, to the employees, customers, and partners we work with.

Of course, as a technologist, I have seen how computing can play a crucial role in improving lives. At home, Zain's speech therapist worked with three high school students to build a Windows app for Zain to control his own music. Zain loves music and has wide-ranging tastes spanning eras, genres, and artists. He likes everything from Leonard Cohen to Abba to Nusrat Fateh Ali Khan and wanted to be able to flip through these artists, filling his room with whatever music suited him at any given moment. The problem was he couldn't control the music on his own—he always had to wait for help, which can be frustrating for him and us. Three high school students studying computer science heard of this problem and wanted to help. Now Zain has a sensor on the side of his wheelchair that he can easily tap his head against to flip through his music collection. What freedom and happiness the empathy of three teenagers has brought to my son.

That same empathy has inspired me at work. Back in our leadership team meeting, to wrap up my discussion, I shared the story of a project we had just completed at Microsoft. Empathy, coupled with new ideas, had helped to create eye-gaze tracking technology, a breakthrough natural user interface that assists people with ALS (also known as Lou Gehrig's disease) and cerebral palsy to have more independence. The idea emerged from the company's first-ever employee hackathon, a hotbed of creativity and dreams. One of the hackathon teams had developed empathy by spending time with Steve Gleason, a former NFL player whose ALS confines him to a wheelchair. Like my son, Steve now uses personal computing technology to improve his daily life. Believe me, I know what this technology will mean for Steve, for millions around the world, and for my son at home.

Our roles on the SLT started to change that day. Each leader was no longer solely employed by Microsoft, they had tapped into a higher calling—to employ Microsoft in pursuit of their personal passions to empower others. It was an emotional and exhausting day, but it set a new tone and put in motion a more unified leadership team. At the end of the day, we all came to the same stark realization: No one leader, no one group, and no one CEO would be the hero of Microsoft's renewal. If there was to be a renewal, it would take all of us and all parts of each of us. Cultural transformation would be slow and trying before it would be rewarding.

—

This is a book about transformation—one that is taking place today inside me and inside of our company, driven by a sense of empathy and a desire to empower others. But most important, it's about the change coming in every life as we witness the most transformative wave of technology yet—one that will include artificial intelligence, mixed reality, and quantum computing. It's about how people, organizations, and societies can and must transform—*hit refresh*—in their persistent quest for new energy, new ideas, relevance, and renewal. At the core, it's about us humans and the unique quality we call empathy, which will become ever more valuable in a world where the torrent of technology will disrupt the status quo like never before. The mystical Austrian poet Rainer Maria Rilke once wrote that "the future enters into us, in order to transform itself in us, long before it happens." As much as elegant computer code for machines, existential poetry can illuminate and instruct us. Speaking to us from another century, Rilke is saying that what lies ahead is very much within us, determined by the course each of us takes today. That course, those decisions, is what I've set out to describe.

In these pages, you will follow three distinct storylines. First, as prologue, I'll share my own transformation moving from India to my new home in America with stops in the heartland, in Silicon Valley, and at a Microsoft then in its ascendancy. Part two focuses on hitting refresh at Microsoft as the unlikely CEO who succeeded Bill Gates and Steve Ballmer. Microsoft's transformation under my leadership is not complete, but I am proud of our progress. In the third and final act, I'll take up the argument that a Fourth Industrial Revolution lies ahead, one in which machine intelligence will rival that of humans. We'll explore some heady questions. What will the role of humans become? Will inequality resolve or worsen? How can governments help? What is the role of multinational corporations and their leaders? How will we hit refresh as a society?

I was excited to write this book, but also a little reluctant. Who really cares about my journey? With only a few years under my belt as Microsoft's CEO, it felt premature to write about how we've succeeded or failed on my watch. We've made a lot of progress since that SLT meeting, but we still have a long way to go. That's also why I'm not interested in writing a memoir. I'll save that for my dotage. But several arguments convinced me to carve out a little time at this stage of my life to write. I felt the tug of responsibility to tell our story from my perspective. It's also a time of enormous social and economic disruption accelerated by technological breakthroughs. The combination of cloud computing, sensors, Big Data, machine learning, and Artificial Intelligence (AI), mixed reality, and robotics foreshadows socioeconomic change ripped from the pages of science fiction. There is a wide and growing spectrum of debate about the implications of this coming wave of intelligent technologies. On the one hand, Pixar's film *WALL-E* paints a portrait of eternal relaxation for humans who rely on robots for the hard work. But on the other, scientists like Stephen Hawking warn of doom.

The most compelling argument was to write for my colleagues—Microsoft's employees—and for our millions of customers and partners. After all, on that cold February day in 2014 when Microsoft's board of directors announced that I would become CEO, I put the company's culture at the top of our agenda. I said that we needed to rediscover the soul of Microsoft, our reason for being. I have come to understand that my primary job is to curate our culture so that one hundred thousand inspired minds—Microsoft's employees—can better shape our future. Books are so often written by leaders looking back on their tenures, not while they're in the fog of war. What if we could share the journey together, the meditations of a sitting CEO in the midst of a massive transformation? Microsoft's roots, its original raison d'être, was to democratize computing, to make it accessible to everyone. "A computer on every desk and in every home" was our original mission. It defined our culture. But much has changed. Most every desk and home now have a computer, and most people have a smartphone. We had succeeded in many ways, but we also were lagging in too many other ways. PC sales had slowed and we were significantly behind in mobile. We were behind in search and we needed to grow again in gaming. We needed to build deeper empathy for our customers and their unarticulated and unmet needs. It was time to hit refresh.

After twenty-two years as an engineer and a leader at Microsoft, I had been more philosophical than anxious about the search process for a new CEO. Even with speculation swirling about who would succeed Steve, quite frankly, my wife, Anu, and I largely ignored the rumors. At home, we were just too busy with taking care of Zain and our two daughters. At work I was very focused on continuing to grow a highly competitive business, the Microsoft Cloud. My attitude was that the board would select the best person. It would be great if it were me. But I would also be equally happy working for someone the board had confidence in. In fact, as part of the interview process one of the board members suggested that if I wanted to be CEO, I needed to be clear that I was hungry for the job. I thought about this and even talked to Steve. He laughed and simply said, "It's too late to be different." It just wouldn't be me to display that kind of personal ambition.

When John Thompson, who at that time was the lead independent director and headed the CEO search, sent me an email on January 24, 2014, asking for time to chat, I was not sure what to make of it. I thought he probably was going to give me an update on where the board was in its decision process. And so, when John called that evening, he first asked me if I was sitting down. I was not. In fact, I was calmly playing with a Kookaburra cricket ball as I usually do when talking on the speakerphone at work. He went on to deliver the news that I was to become the new CEO of Microsoft. It took a couple of minutes to digest his message. I said that I was honored, humbled, and excited. They were unplanned words, but they perfectly captured how I felt. Weeks later, I told media outlets that we needed to focus more clearly, move faster, and continue to transform our culture and business. But behind the scenes, I knew that to lead effectively I needed to get some things square in my own mind—and, ultimately, in the minds of everyone who works at Microsoft. Why does Microsoft exist? And why do I exist in this new role? These are questions everyone in every organization should ask themselves. I worried that failing to ask these questions, and truly answer them, risked perpetuating earlier mistakes and, worse, not being honest. Every person, organization, and even society reaches a point at which they owe it to themselves to hit refresh—to reenergize, renew, reframe, and rethink their purpose. If only it were as easy as punching that little refresh button on your browser. Sure, in this age of continuous updates and always-on technologies, hitting refresh may sound quaint, but still when it's done right, when people and cultures re-create and refresh, a renaissance can be the result. Sports franchises do it. Apple did it. Detroit is doing it. One day ascending companies like Facebook will stop growing, and they will have to do it too.

And so let me start at the beginning—my own story. I mean, what kind of CEO asks such existential questions as why do we exist in the first place? Why are concepts like culture, ideas, and empathy so important to me? Well, my father was a civil servant with Marxist leanings and my mother was a Sanskrit scholar. While there is much I learned from my father, including intellectual curiosity and a love of history, I was always my mother's son. She cared deeply about my being happy, confident, and living in the moment without regrets. She worked hard both at home and in the college classroom where she taught the ancient language, literature, and philosophy of India. And she created a home full of joy.

Even so, my earliest memories are of my mom struggling to continue her profession and to make the marriage work. She was the constant, steadying force in my life, and my father was larger than life. He nearly immigrated to the United States, a faraway place that represented opportunity, on a Fulbright fellowship to pursue a PhD in economics. But those plans were suddenly and understandably shelved when he was selected to join the Indian Administrative Service (IAS). It was early 1960s, and Jawaharlal Nehru was India's first prime minister following Gandhi's historic movement, which had achieved independence from Great Britain. For that generation entering the civil service and being part of the birth of a new nation was a true dream come true. The IAS was essentially a remnant of the old Raj system left by the British to govern after the UK turned over control of the country in 1947. Only about a hundred young professionals per year were selected for the IAS, and so at a very young age my father was administering a district with millions of people. Throughout my childhood, he was posted in many districts across the state of Andhra Pradesh in India. I remember moving from place to place, growing up in the sixties and early seventies in old colonial buildings in the middle of nowhere with lots of time and space, and in a country being transformed.

My mom did her level best during all these disruptions to maintain her teaching career, raise me, and be a loving wife. When I was about six, my five-month-old sister died. It had a huge impact on me and our family. Mom had to give up working after that. I think my sister's death was the last straw. Losing her, combined with raising me and working to maintain a career while my father was working in faraway places was just too much. She never complained to me at all about it, but I reflect on her story quite a bit, especially in the context of today's diversity conversations across the technology industry. Like anyone, she wanted to, and deserved to, have it all. But the culture of her workplace, coupled with the social norms of Indian society at the time, didn't make it possible for her to balance family life with her professional passions.

Among the children of IAS fathers, it was a rat race. For some of the IAS dads, simply passing the grueling entrance test meant they were set for life. It was the last test they would ever have to take. But my father believed passing the IAS exam was merely the entry point to being able to take even more important exams. He was a quintessential lifelong learner. But unlike most of my peers at that time, whose high-achieving parents applied tremendous pressure to achieve, I didn't face any of that. My mom was just the opposite of a tiger mom. She never pressured me to do anything other than just be happy.

That suited me just fine. As a kid, I couldn't have cared less about pretty much anything, except for the sport of cricket. One time, my father hung a poster of Karl Marx in my bedroom; in response, my mother hung one of Lakshmi, the Indian goddess of plentitude and contentment. Their contrasting messages were clear: My father wanted intellectual ambition for me, while my mother wanted me to be happy versus being captive to any dogma. My reaction? The only poster I really wanted was one of my cricketing hero, the Hyderabadi great, M. L. Jaisimha, famous for his boyish good looks and graceful style, on and off the field.

Looking back, I have been influenced by both my father's enthusiasm for intellectual engagement and my mother's dream of a balanced life for me. And even today, cricket remains my passion. Nowhere is the intensity for cricket greater than in India, even if the game was invented in England. I was good enough to play for my school in Hyderabad, a place that had a lot of cricket tradition and zeal. I was an off-spin bowler, which in baseball would be the equivalent to a pitcher with a sharp breaking curveball. Cricket attracts an estimated 2.5 billion fans globally, compared with just half a billion baseball fans. Both are beautiful sports with passionate fans and a body of literature brimming with the grace, excitement, and complexities of competition. In his novel, *Netherland*, Joseph O'Neill describes the beauty of the game, its eleven players converging in unison toward the batsman and then returning again and again to their starting point, "a repetition or pulmonary rhythm, as if the field breathed through its luminous visitors." I think of that metaphor of the cricket team now as a CEO when reflecting on the culture we need in order to be successful.

I had attended schools in many parts of India—Srikakulam, Tirupati, Mussoorie, Delhi, and Hyderabad. Each left its mark and has remained with me. Mussoorie, for example, is a northern Indian city tucked into the foothills of the Himalayas, around six thousand feet of elevation. Every time I see Mount Rainier from my home in Bellevue, I am always reminded of the mountains of childhood—Nanda Devi and Bandarpunch. I attended kindergarten at the Convent of Jesus and Mary. It is the oldest school for girls in India but they let boys attend kindergarten. By age fifteen, we had stopped moving and I entered Hyderabad Public School, which boarded students from all over India. I'm thankful for all the moves—they helped me adjust quickly to new situations—but going to Hyderabad was truly formative. In the 1970s, Hyderabad was out of the way, not at all the metropolis of 6.8 million people it is today. I really didn't know or care about the world west of Bombay on the Arabian Sea, but attending boarding school at HPS was the best break I had in my life.

At HPS I belonged to the Nalanda, or blues house, which was named for an ancient Buddhist university. The whole school was multicultural: Muslims, Hindus, Christians, Sikhs all living and studying together. The school was attended by members of the elite as well as by tribal kids who had come from the interior districts on scholarships. The chief minister's son attended HPS alongside the children of Bollywood actors. In my dorm there were kids from every part of the Indian economic strata. It was an amazingly equalizing force—a moment in time worth remembering.

The list of alumni today speaks to this success. Shantanu Narayen, the CEO of Adobe; Ajay Singh Banga, the CEO of MasterCard; Syed B. Ali, head of Cavium Networks; Prem Watsa, founder of Fairfax Financial Holdings in Toronto; parliament leaders, film stars, athletes, academics, and writers—all came from this small, out-of-the-way school. I was not academically great and nor was the school known to push academics. If you liked to study physics, you studied physics. If you felt like, oh, science was too boring and you wanted to study history, you studied history. There wasn't that intense peer pressure to follow a particular path.

After a few years at HPS my dad went to work at the United Nations in Bangkok. He wasn't too fond of my laid-back attitude. He said, "I'm going to pull you out and you should come do your eleventh and twelfth in some international school in Bangkok." I said no chance. And so I just stuck to Hyderabad. Everybody was thinking, "Are you crazy, why would you do that?" But there was no way I was leaving. Cricket was a major part of my life at that time. Attending that school gave me some of my greatest memories, and a lot of confidence.

By twelfth grade if you had asked me about my dream it was to attend a small college, play cricket for Hyderabad, and eventually work for a bank. That was it. Being an engineer and going to the West never occurred to me. My mom was happy with those plans. "That's fantastic, son!" But my dad really forced the issue.

He said, "Look, you've got to get out of Hyderabad. Otherwise you'll ruin yourself." It was good advice then, but few could predict that Hyderabad would become the technological hub it is today. It was hard to break from my circle of friends, but Dad was right. I was being provincial with my ambitions. I needed some perspective. Cricket was my passion, but computers were a close second. When I was fifteen, my father brought me a Sinclair ZX Spectrum computer kit from Bangkok. Its Z80 CPU had been developed in the mid-seventies by an engineer who left Intel, where he had been working on the 8080 microprocessor, which ironically was the chip Bill Gates and Paul Allen used to write the original version of Microsoft BASIC. The ZX Spectrum inspired me to think about software, engineering, and even the idea that personal computing technologies could be democratizing. If a kid in nowhere India could learn to program, surely anyone could.

I flunked the Indian Institutes of Technology (IIT) entrance exam, the holy grail of all things academic for middle-class kids growing up in India at that time. My father, who never met an entrance test he did not pass, was more amused than annoyed. But, luckily, I had two other options to pursue engineering. I had gotten into mechanical engineering at Birla Institute of Technology in Mesra and electrical engineering (EE) at Manipal Institute of Technology. I chose Manipal based on a hunch that pursuing EE was going to get me closer to computers and software. And fortuitously the hunch was right. Academically it put me on a pathway that would lead to Silicon Valley and eventually to Microsoft. The friends I made in college were entrepreneurial, driven, and ambitious. I learned from many of them. In fact, years later I rented a house in Sunnyvale, California, with eight of my classmates from Manipal and re-created our dorm-room experience from college. Athletically, though, Manipal left a lot to be desired. Playing cricket was no longer my central passion. I played one match for my college team and hung up my gear. Computers took cricket's place and became number one in my life. At Manipal I trained in microelectronics—integrated circuits and the first principles of making computers.

I didn't really have a specific plan for what I'd do after finishing my electrical engineering degree. There is much to be said for my mother's philosophy of life, which influenced how I thought about my own future and opportunities. She always believed in doing your thing, and at your pace. Pace comes when you do *your* thing. So long as you enjoy it, do it mindfully and well, and have an honest purpose behind it, life won't fail you. That has stood me in good stead all my life. After graduation, I had an opportunity to attend a prestigious industrial engineering institute in Bombay. I had also applied to a few colleges in the United States. In those days, the student visa was bit of a crapshoot, and frankly I was hoping it would be rejected. I never wanted to leave India. But as fate would have it, I got my visa and was again faced with some choices—whether to stay in India and do a master's degree in industrial engineering or go to the University of Wisconsin at Milwaukee for a master's degree in electrical engineering. A very dear friend from HPS was attending Wisconsin studying computer science, and so my decision was made. I entered the master's in computer science program at Wisconsin. And I'm glad I did because it was a small department with professors who were invested in their students. I'm particularly thankful to then department chair Dr. Vairavan and my master's advisor Professor Hosseini for instilling in me the confidence not to pursue what was easy, but to tackle the biggest and hardest problems in computer science.

If someone had asked me to point to Milwaukee on a map I could not have done it. But on my twenty-first birthday, in 1988, I flew from New Delhi to Chicago O'Hare Airport. From there a friend drove me to campus and dropped me off. What I remember was the quiet. Everything was quiet. Milwaukee was just stunning, pristine. I thought, god, this place is heaven on earth. It was summer. It was beautiful, and my life in the United States was just beginning.

Summer became winter and the cold of Wisconsin is something to behold if you've come from southern India. I was a smoker at the time and all smokers had to stand outside. There were a number of us from various parts of the world. The Indian students couldn't stand the cold so we quit smoking. Then my Chinese friends quit. But the Russians were unaffected by winter's chill, and they just kept on puffing away.

Sure, I would get homesick, like any kid, but America could not have been more welcoming. I don't think my story would be possible anywhere else, and I am proud today to call myself an American citizen. Looking back, though, I suppose my story may sound a little programmatic. The son of an Indian civil servant studies hard, gets an engineering degree, immigrates to the United States, and makes it in tech. But it wasn't that simple. Unlike the stereotype, I was actually not academically that great. I didn't go to the elite Indian Institutes of Technology (IITs) that have become synonymous with building Silicon Valley. Only in America would someone like me get the chance to prove himself rather than be typecast based on the school I attended. I suppose that was true for earlier waves of immigration as well and will be just as true for new generations of immigrants.

Like many others, it was my great fortune to benefit from the convergence of several tectonic movements: India's independence from British rule, the American civil rights movement, which changed immigration policy in the United States, and the global tech boom. Indian independence led to large investments in education for Indian citizens like me. In the United States, the 1965 Immigration and Naturalization Act abolished the nation-of-origin quota and made it possible for skilled workers to come to America and contribute. Before that, only about a hundred Indians were allowed to immigrate each year. Writing for *The New York Times* on the fiftieth anniversary of the immigration act, historian Ted Widmer noted that nearly 59 million people came to the United States as a result of the act. But the influx was not unrestrained. The act created preferences for those with technical training and those with family members already in the States. Unknowingly, I was the recipient of this great gift. These movements enabled me to show up in the United States with software skills just before the tech boom of the 1990s. Talk about hitting the lottery.

During the first semester at Wisconsin, I took image processing, a computer architecture class, and LISP, one of the oldest computer programming languages. The first set of assignments were just huge programming projects. I'd written a little bit of code but I was not a proficient coder by any stretch. I know the stereotype in America is that the Indians who immigrate are born to code, but we all start somewhere. The assignments were, basically, here it is, go write a bunch of code. It was tough and I had to pick it up quickly. Once I did, it was awesome. I understood pretty early on that the microcomputer was going to shape the world. Initially I thought it might be all about building chips. Most of my college friends all went on to specialize in chip design and work at places with real impact like Mentor Graphics, Synopsys, and Juniper.

I became particularly interested in a theoretical aspect of computer science that was, at its heart, designed to make fast decisions in an atmosphere of great uncertainty and finite time. My focus was a computer science puzzle known as graph coloring. No, I wasn't coloring graphs with crayons. Graph coloring is part of computational complexity theory in which you must assign labels, traditionally called colors, to elements of a graph within certain constraints. Think of it this way: Imagine coloring the U.S. map so that no state sharing a common border receives the same color. What is the minimal number of colors you would need to accomplish this task? My master's thesis was about developing the best heuristics to accomplish complex graph coloring in nondeterministic polynomial time, or NP-complete. In other words, how can I solve a problem that has limitless possibilities in a way that is fast and good but not always optimal? Do we solve this as best we can right now, or work forever for the best solution?

Theoretical computer science really grabbed me because it showed the limits to what today's computers can do. It led me to become fascinated by mathematicians and computer scientists John Von Neumann and Alan Turing, and by quantum computing, which I will write about later as we look ahead to artificial intelligence and machine learning. And, if you think about it, this was great training for a CEO—nimbly managing within constraints.

I completed my master's in computer science at Wisconsin and even managed to work for what Microsoft would now call an independent software vendor (ISV). I was building apps for Oracle databases while finishing my master's thesis. I was good at relational algebra and became proficient with databases and structured query language (SQL) programming. This was the era where technology was changing from character or text mode on UNIX workstations to graphical user interfaces like Windows. It was early 1990 and I didn't even really think about Microsoft at that time because we never used PCs. My focus was on more powerful workstations.

In fact, I left Milwaukee in 1990 for my first job in Silicon Valley at Sun Microsystems. Sun was the king of workstations, a market Microsoft had in its crosshairs. Sun had an amazing collection of talent, including its founders Scott McNealy and Bill Joy, as well as James Gosling, the inventor of Java, and Eric Schmidt, our VP for software development who went on to run Novell and then Google.

My two years at Sun were a time of great transition in the computer business as Sun looked longingly at Microsoft's Windows graphical user interface, and Microsoft looked longingly at Sun's beautiful, powerful 32-bit workstations and operating systems. Again, I happened to be at the right place at the right time. Sun asked me to work on desktop software like their email tool. I was later sent to Cambridge, Massachusetts, to work for several months with Lotus to port their spreadsheet software to Sun workstations. Then I started to notice something alarming. Every couple of months, Sun wanted to adopt a new graphical user interface (GUI) strategy. That meant I had to rework my programs constantly, and their explanations made less and less sense. I realized that despite its phenomenal leadership and capability, it had a hard time building and sticking with a cogent software strategy.

By 1992, I was again at a crossroads in my life. I wanted to work on software that would change the world. I also wanted to return to graduate school for my MBA. And I missed Anu, whom I intended to marry and bring to the United States. She was finishing her degree in architecture back in Manipal, and we began to plan for her to join me in America.

Like all the times before, there was no master plan, but a call from Redmond, Washington, one afternoon would create a new, unexpected opportunity. It was time to hit refresh again.

———

On a cool, November day in the Pacific Northwest, I first set foot on the Microsoft campus and entered an unremarkable corporate office unimaginatively named Building 22. Shrouded by towering Douglas firs, it remains even today barely visible from the adjacent state route 520, known for its floating bridge connecting Seattle to Redmond. The year was 1992. Microsoft's stock was just beginning an epic rise, though its founders, Bill Gates and Paul Allen, could still walk down the street unrecognized. Windows 3.1 had just been released, setting the stage for Windows 95 and the grandest consumer technology product launch yet. Sony introduced the CD-ROM, and the first website was launched, though it would be two more years before the Internet would become a tidal wave. TCI introduced digital cable and the FCC approved digital radio. On a chart, PC sales at this time show the start of a meteoric ascent. Looking back now, I couldn't have timed my entrance any better. The resources, the talent and the vision were there to compete and to lead the industry. My journey to Redmond had taken me from my home in India to Wisconsin for graduate school to the Silicon Valley to work for Sun. Over the summer I had been recruited to join Microsoft as a twenty-five-year-old evangelist for Windows NT, a 32-bit operating system that was designed to extend the company's popular consumer program into much more powerful business systems. A few years later NT would become the backbone of future Windows versions. Even today's generation of Windows, Windows 10, builds on the original NT architecture. I had heard of NT while working at Sun but had never used it. A colleague had attended a Microsoft conference where they showed off NT to developers. He came back and told me about the product. I thought, wow, this is going to get serious. I wanted to be in a place that would have real impact. The guys who had recruited me to Microsoft, Richard Tait and Jeff Teper, said they needed someone who understood UNIX and 32-bit operating systems. I was a little unsure. What I really wanted to do was go to business school. I knew that management would complement my engineering training, and I had been thinking about a switch to investment banking. I had gotten into the full-time program at University of Chicago, but Teper said, "You should just join us straightaway." I decided to do both. I was able to switch my admission to the part-time program at Chicago, but then never told anyone that I was flying to Chicago for weekends. I finished my MBA in two years and was glad I did. During the week my job was to fly all over the country—lugging these enormous Compaq computers—to meet with customers, usually chief information officers at places like Georgia Pacific or Mobil, to convince them that our new, more robust operating system for business was superior to the others and convert them. And at school I learned more math by taking high-level finance classes in Chicago than in my engineering coursework. The classes I took with Steven Kaplan, Marvin Zonis, and many other storied faculty at the university on strategy, finance, and leadership influenced my thinking and intellectual pursuits long after I completed the MBA. It was an exciting time to be at Microsoft. Not long after joining I met Steve Ballmer for the first time. He stopped by my office to give me one of his very expressive high fives for leaving Sun and joining Microsoft. It was the first of what would be many interesting and enjoyable conversations with Steve over the years. There was a true sense of mission and energy at the company then. The sky was the limit.

—

Within a few years my work on Windows NT landed me in a new advanced technology group, founded by Renaissance man Nathan Myhrvold. Along with Rick Rashid, Craig Mundie, and others, Microsoft was assembling the greatest technology IQ since Xerox PARC, the famed Silicon Valley center for innovation. I was humbled when asked to join the group as a product manager on a project code-named Tiger Server, which was a major investment in building a video-on-demand (VOD) service. It would be years before cable companies would deliver the technology and business model to support VOD, and years before Netflix made video streaming mainstream. Fortunately, I lived right next to the Microsoft campus, the endpoint for all of this amazing broadband infrastructure that made our VOD pilot possible. So in 1994, long before it was commercially available, I had video-on-demand while sitting in my little apartment. We only had about fifteen movies but I remember watching them over and over again. Even as our team planned to launch our Tiger server over a fully switched asynchronous transfer mode (ATM) network to the home, we saw our idea become obsolete virtually overnight with the birth of the Internet.

—

While my mind was fully engaged, my heart was distracted. Anu and I had decided to marry when I made a trip back to India just before joining Microsoft. I had known Anu all of my life. Her dad and my father had joined the IAS together and we were family friends. In fact, Anu's dad and I shared a passion for talking endlessly about cricket, something we continue to this day. He had played for his school and college, captaining both teams. When exactly I fell in love with Anu is what computer scientists would call an NP complete question. I can come up with many times and places but there is no one answer. In other words, it's complex. Our families were close. Our social circles were the same. As kids we had played together. We overlapped in school and college. Our beloved family dog came from Anu's family dog's litter. But once I moved to the United States, I lost touch with her. When I went back to India for a visit, we saw each other again. She was in her final year of architecture at Manipal and enjoying an internship in New Delhi. Our two families met for dinner one evening, and that night, more than ever, I was convinced that she was the one. We shared the same values, the same outlook on the world, and dreamed of similar futures. In many ways, her family was already mine and mine hers. The next day, I persuaded her to take me to an optician where I needed to have my glasses repaired. After the appointment, we walked and talked for hours in the neighboring Lodi Gardens, an ancient architectural site that today is popular with tourists. Anu, a student of architecture, loved all the historical monuments that dotted Delhi, and for days afterward we explored them together. I had visited them all before as a kid. But this was different. We stopped for lunch on Pandara Road, enjoyed plays in the National Institute of Drama, and shopped in the bookstores of Khan Market. We had fallen in love. It was in the lush Lodi Gardens that one October afternoon in 1992 I proposed and, thankfully for me, Anu said yes. We walked back to Anu's place on Humayun Road and broke the news to Anu's mom. We were married just two months later, in December. It was a happy time, but the complications of immigration would soon prove a challenge.

———

Anu was in the last year of her architecture degree and the plan was for her to complete the remaining course and join me in Redmond. In the summer of 1993, Anu applied for a visa to join me during her final vacation before finishing school. But her visa application was rejected because she was married to a permanent resident. Anu's father sought an appointment with the U.S. consul general in New Delhi and argued with him that the U.S. visa rules were not consistent with the family values that the United States stood for. The combination of his persuasiveness and the kindness of the U.S. consul general led to Anu getting a short-term tourist visa—a rare exception. After her vacation, she returned to India and college to complete her degree. It was now clear to us that Anu's return to the United States would be very difficult given the visa waitlist for spouses of permanent residents. Microsoft had an immigration lawyer who told me it would take five or more years to get Anu into the country under existing rules. I contemplated quitting Microsoft and returning to India. But our lawyer, Ira Rubinstein, said something interesting. "Hey, maybe you should give up your green card and go back to an H1B." He was suggesting that I give up permanent residency and instead reapply for temporary professional worker status. If you've seen the Gerard Depardieu film *Green Card*, you know the comedic lengths people will go to to obtain permanent residency in the United States. So why would I give up the coveted green card for temporary status? Well, the H1B enables spouses to come to the United States while their husbands and wives are working here. Such is the perverse logic of this immigration law. There was nothing I could do about it. Anu was my priority. And that made my decision a simple one. I went back to the U.S. embassy in Delhi in June of 1994, past the enormous lines of people hoping to get a visa, and told a clerk that I wanted to give back my green card and apply for an H1B. He was dumbfounded. "Why?" he asked. I said something about the crazy immigration policy, he shook his head and pushed a new form to me. "Fill this out." The next morning, I returned to apply for an H1B application. Miraculously, it all worked. Anu joined me (for good) in Seattle, where we would start a family and build a life together. What I didn't expect was the instant notoriety around campus. "Hey, there goes the guy who gave up his green card." Every other day someone would call me and ask for advice. Much later, one of my colleagues, Kunal Bahl, did quit Microsoft when his H1B ran out and his green card had not yet arrived. He returned to India and then founded Snapdeal, which today is worth more than $1 billion and employs five thousand people. Ironically, online, cloud-based companies like Snapdeal would play an important role in my future and that of Microsoft. And the lessons I learned in my former country continue to shape my present.

CHAPTER 2

Learning to Lead

Seeing the Cloud Through Our Windows

I am obsessed with cricket. No matter where I am, this beautiful game is always in the back of my mind. The joy, the memories, the drama, the complexities, and the ups and downs—the infinite possibilities.

For those of you unfamiliar with cricket, it is an international sport played on a large green oval in the summer and early fall. Its popularity is strongest among the current and former nations of the British Commonwealth. Like baseball, in cricket a ball is hurled at a batter who endeavors to strike the ball and score as many runs as possible. The pitcher is a bowler, the batter is a batsman, the diamond is a wicket, and the fielders try to get the batsman out. Yes, there are forms of a match that can stretch on for days, but then in baseball teams compete to win 3-, 5-, and even 7-game series. Both sports are endlessly complex, but suffice it to say that the team with the most runs wins. This is not the book to describe the ins and outs of cricket, but it is a book that cannot avoid the metaphor of cricket and business.

Like most South Asians, I somehow fell in love with this most English of games on the dusty matting wickets of the Deccan Plateau in southern India.

There, on those fields, I learned a lot about myself—succeeding and failing as a bowler, a batsman, and a fielder. Even today I catch myself reflecting on the nuances within the cricket rulebook and the inherent grace of a team of eleven working together as one unit.

During the early years of my life when my father's work as a civil servant took us to the district headquarters of Andhra Pradesh and the hills of Mussoorie in what is now Uttarakhand, cricket was not the phenomenon it is now. Today the Indian Premier League sells its ten-year television rights for billions. But back then it became a phenomenon for me, when at the age of eight, we moved to Hyderabad. We stayed in a rented house in the Somajiguda neighborhood, and our landlord, Mr. Ali, was a gracious and proud Hyderabadi, who wore his Osmania University cricket cap while working in his auto shop. He was full of stories about all the great Hyderabadi cricketers of the 1960s. He once took me to watch a first-class match between Hyderabad and Bombay (today's Mumbai). It was my first time in the great cricket stadium Fateh Maidan. I was completely smitten that day with all the glamour of cricket. The athletes, M. L. Jaisimha, Abbas Ali Baig, Abid Ali, and Mumtaz Hussain, became my heroes. The Bombay side had Sunil Gavasker and Ashok Mankad, among many other stars. I don't recall any of them making much of an impression, even though they beat Hyderabad handily. I was in awe of M. L. Jaisimha's on-field presence—his fashionable upturned collar and distinctive gait. To this day I remember Mr. Ali's descriptions of Mumtaz Hussain's "mystery ball," and watching Abid Ali charging down the wicket to a medium pacer.

Soon my dad was again transferred in his job, and I moved to attend school in Delhi. There I watched my first Test match at Feroz Shah Kotla. It was a match between India and England. Watching these two sides play left an indelible impression. I remember the English batsman Dennis Amiss and bowler John Lever combined to destroy India by an inning, leaving me distraught for weeks. Amiss hit a double hundred, and Lever, playing in his first Test match, bowled medium pace through that long afternoon, and the ball was swinging for him like I'd never seen before. Suddenly all the Indian players were back in the hut.

When I was ten I returned to Hyderabad, and for the next six years I truly and surely fell in love with cricket as a player for Hyderabad Public School (HPS). In fact, Mr. Jaisimha's two children attended my school, and as a result we were surrounded by cricket glamour, tradition, and obsession. In those days, everyone was talking about the two India School players from HPS. One of them was Saad Bin Jung (who also happened to be the famous Indian cricket captain, Tiger Pataudi's, nephew). Still in school, he went on to smash a hundred runs against a touring West Indian side while playing for South Zone, representing our region of southern India. I began playing on the B team and graduated to the senior team, which played in the A leagues of Hyderabad. We were the only school team to play in the A leagues as the other teams were sponsored by banks and miscellaneous companies. Ranji Trophy players would turn up in these league games, and all that intrigue made for intense competition.

What excited me then about cricket is what still excites me today, even living in a non-cricketing country (though, the United States over a hundred years ago did periodically host Australian and English sides). Cricket for me is like a wondrous Russian novel with plots and subplots played out over the course of multiple acts. In the end, one brilliant knock, or three deftly bowled balls, can change the complexion of a game.

There are three stories from my all-too-brief cricketing past that speak very directly to business and leadership principles I use even today as a CEO.

The first principle is to compete vigorously and with passion in the face of uncertainty and intimidation. In my school cricketing days, we played a team one summer that had several Australian players. During the match, our PE teacher, who acted as a sort of general manager for the team, noticed that we were admiring the Aussies' play. In fact, we were more than a little intimidated by them. We had never played against foreign players, and Australia of course loomed large in the national cricket psyche. I now recognize our teacher and general manager as very much like an American football coach—loud and very competitive. He was having none of our admiration and intimidation. He began by yelling at the captain to get more aggressive. I was a bowler and a terrible fielder but he positioned me at forward short leg, right beside the powerful Australian batsmen. I would have been happy standing far away, but he put me right next to the action. In time, with new energy and new focus, we transformed into a competitive team. It showed me that you must always have respect for your competitor, but don't be in awe. Go and compete.

On reflection, a second principle is simply the importance of putting your team first, ahead of your personal statistics and recognition. One of my teams had a brilliant fast bowler. He was one of the most promising young cricketers in the land. He became even greater after attending a U-19 South Zone coaching clinic. His pace and accuracy were just brilliant. As a tail-end batsman myself, being in the nets (similar to baseball batting cages) against this guy was tough. But he had a self-destructive mindset. During one game our captain decided to replace him with another bowler. Soon, the new bowler coaxed the opposing batsman to mis-hit a ball skyward, an easy catch for our cantankerous teammate now at mid-off, a fielding position twenty-five to thirty yards from the batsman. Rather than take a simple catch, he plunged both hands deep into his pockets and watched passively as the ball fell right in front of him. He was a star player, and we looked on in complete disbelief. The lesson? One brilliant character who does not put team first can destroy the entire team.

There are of course many lessons and principles one can take from cricket, but for me a third is the central importance of leadership. Looking back, I remember one particular match in which my off-spin bowling was getting hammered by the opponents. I was serving up very ordinary stuff. Our team captain in retrospect showed me what real leadership looks like. When my over had ended (that is, when I had thrown six balls), he replaced me with himself even though he was a better batsman than bowler. He quickly took the wicket—the batsman was out. Customarily taking a wicket that efficiently would argue for him remaining in as a bowler. But instead, he immediately handed the ball back to me and I took seven wickets of my own. Why did he do it? I surmised he wanted me to get my confidence back. It was early in the season and he needed me to be effective all year. He was an empathetic leader, and he knew that if I lost my confidence it would be hard to get it back. That is what leadership is about. It's about bringing out the best in everyone. It was a subtle, important leadership lesson about when to intervene and when to build the confidence of an individual and a team. I think that is perhaps the number one thing that leaders have to do: to bolster the confidence of the people you're leading. That team captain went on to play many years of prestigious Ranji Trophy competition, and he taught me a very valuable lesson.

Those early lessons from cricket shaped my leadership style, as have my experiences as a husband, a father, a young Microsoft engineer thrilled to be part of our company's visionary ascent, and later as an executive charged with building new businesses. My approach has never been to conduct business as usual. Instead it's been to focus on culture and imagine what's possible. The culmination of these experiences has provided the raw material for the transformation we are undergoing today—a set of principles based on the alchemy of purpose, innovation, and empathy.

—

The arrival of our son, Zain, in August 1996 had been a water-shed moment in Anu's and my life together. His suffering from asphyxia *in utero*, had changed our lives in ways we had not anticipated. We came to understand life's problems as something that cannot always be solved in the manner we want. Instead we had to learn to cope. When Zain came home from the intensive care unit (ICU), Anu internalized this understanding immediately. There were multiple therapies to be administered to him every day, not to mention quite a few surgeries he needed that called for strenuous follow-up care after nerve-racking ICU stays. All this entailed Anu lovingly placing him in the infant car seat and driving him, day after day, from the early hours of the day, from therapist to therapist, not to mention frequent visits to the ICU unit at Seattle Children's Hospital. Children's became a second home for our family as Zain's medical file grew to over a foot high. We are today, as we always have been, so indebted to the staff at Children's who have loved and cared for Zain throughout his life from infancy to young adulthood.

During one ICU visit, after I took on my new role as CEO, I looked around Zain's room, filled with the soft buzzing and beeping of medical technology, and saw things differently. I noticed just how many of the devices ran on Windows and how they were increasingly connected to the cloud, that network of massive data storage and computational power that is now a fundamental part of the technology applications we take for granted today. It was a stark reminder that our work at Microsoft transcended business, that it made life itself possible for a fragile young boy. It also brought a new level of gravity to the looming decisions back at the office on our cloud and Windows 10 upgrades. We'd better get this right, I remember thinking to myself.

My son's condition requires that I draw daily upon the very same passion for ideas and empathy that I learned from my parents. And I do this both at home and at work. Whether I am meeting with people in Latin America, the Middle East, or one of the inner cities of America, I am always searching to understand people's thoughts, feelings, and ideas. Being an empathetic father, and bringing that desire to discover what is at the core, the soul, makes me a better leader.

But it is impossible to be an empathetic leader sitting in an office behind a computer screen all day. An empathetic leader needs to be out in the world, meeting people where they live and seeing how the technology we create affects their daily activities. So many people around the world today depend on mobile and cloud technologies without knowing it. Hospitals, schools, businesses, and researchers rely on what's referred to as the "public cloud"—an array of large-scale, privacy-protected computers and data services accessible over a public network. Cloud computing makes it possible to analyze vast quantities of data to produce specific insights and intelligence, converting guesswork and speculation into predictive power. It has the power to transform lives, companies, and societies.

Traveling the globe as CEO, I've seen example after example of this interplay between empathy and technology.

Both in the state where I was born and the state in which I

now live, schools use the power of cloud computing to analyze large amounts of data to uncover insights that can improve dropout rates. In Andhra Pradesh in India, and in Tacoma, Washington, too many kids drop out of school. The problem is lack of resources, not lack of ambition. Cloud technology is helping improve outcomes for kids and families as intelligence from cloud data is now predicting which students are most likely to drop out of school so that resources can be focused on providing them the help they need.

Thanks to mobile and cloud technologies, a startup in Kenya has built a solar grid that people living on less than two dollars a day can lease to have safe, low-cost lighting and efficient cookstoves, replacing polluting and dangerous kerosene power. It's an ingenious plan because the startup can effectively create a credit rating, a byproduct of the service, which, for the first time, gives these Kenyans access to capital. This innovative mobile phone payment system enables customers living in Kenya's sprawling slums to make forty-cent daily payments for solar light, which in turn generates data that establishes a credit history to finance other needs.

A university in Greece, leveraging cloud data, is working with firefighters in that country to predict and prevent massive wildfires like the one in 2007 that killed eighty-four people and burned 670,000 acres. Firefighters are now armed with intelligence on the rate of the fire's spread, intensity, movement of the perimeter, proximity to water supply, and microclimate weather forecasts from remote sensors, enabling them to catch fires early, saving lives and property.

In Sweden, researchers are using cloud technologies to ensure that children are screened earlier and more accurately for dyslexia, a reading disorder that impacts educational outcomes for millions. Eye movement data analyzed at schools today can be compared with a data set from those diagnosed with dyslexia thirty years ago. Diagnostic accuracy rates have increased from 70 to 95 percent, and the time to get a diagnosis has decreased from three years to three minutes. This means students, parents, and schools are prepared earlier and struggle less.

In Japan, crowd-sourced data collected from hundreds of sensors nationwide helped the public monitor radiation from the Fukushima nuclear plant to reduce risks to food quality and transportation. The 13 million measurements from five hundred remote sensors generated a heat map that alerted authorities to threats to local rice production.

And in Nepal, after the devastating earthquake there in April 2015, disaster relief workers from the United Nations used the public cloud to collect and analyze massive amounts of data about schools, hospitals, and homes to speed up access to compensatory entitlements, relief packages, and other assistance.

Today it's hard to imagine devices that are not connected to the cloud. Consumer applications like O365, LinkedIn, Uber, and Facebook all live in the cloud. There's a great scene in Sylvester Stallone's *Creed*, the latest of his *Rocky* movie series. The champ jots down on a piece of paper a workout regimen for his protégé, who quickly snaps a photo of it on his smartphone. As the kid jogs away, Rocky yells, "Don't you want the paper?"

"I got it right here, it's already up in the cloud," the kid replies.

The aging Rocky looks skyward. "What cloud? What cloud?" Rocky may not know about the cloud, but millions of others rely on it.

Microsoft is at the leading edge of today's game-changing cloud-based technologies. But just a few years ago, that outcome seemed very doubtful.

By 2008, storm clouds were gathering over Microsoft. PC shipments, the financial lifeblood of Microsoft, had leveled off. Meanwhile sales of Apple and Google smartphones and tablets were on the rise, producing growing revenues from search and online advertising that Microsoft hadn't matched. Meanwhile, Amazon had quietly launched Amazon Web Services (AWS), establishing itself for years to come as a leader in the lucrative, rapidly growing cloud services business.

The logic behind the advent of the cloud was simple and compelling. The PC Revolution of the 1980s, led by Microsoft, Intel, Apple, and others, had made computing accessible to homes and offices around the world. The 1990s had ushered in the client/server era to meet the needs of millions of users who wanted to share data over networks rather than on floppy disks. But the cost of maintaining servers in an ever-growing sea of data—and the advent of businesses like Amazon, Office 365, Google, and Facebook—simply outpaced the ability for servers to keep up. The emergence of cloud services fundamentally shifted the economics of computing. It standardized and pooled computing resources and automated maintenance tasks once done manually. It allowed for elastic scaling up or down on a self-service, pay-as-you-go basis. Cloud providers invested in enormous data centers around the world and then rented them out at a lower cost per user. This was the Cloud Revolution.

Amazon was one of the first to cash in with AWS. They figured out early on that the same cloud infrastructure they used to sell books, movies, and other retail items could be rented, like a time-share, to other businesses and startups at a much lower price than it would take for each company to build its own cloud. By June 2008, Amazon already had 180,000 developers building applications and services for their cloud platform. Microsoft did not yet have a commercially viable cloud platform.

All of this spelled trouble for Microsoft. Even before the Great Recession of 2008, our stock had begun a downward slide. In a long-planned move, Bill Gates left the company that year to focus on the Bill & Melinda Gates Foundation. But others were leaving, too. Among them, Kevin Johnson, president of the Windows and online services business, announced he would leave to become CEO of Juniper Networks. In their letter to shareholders that year, Bill and Steve Ballmer noted that Ray Ozzie, creator of Lotus Notes, had been named the company's new Chief Software Architect (Bill's old title), reflecting the fact that a new generation of leaders was stepping up in areas like online advertising and search.

There was no mention of the cloud in that year's shareholder letter, but, to his credit, Steve had a game plan and a wider view of the playing field. Always a bold, courageous, and famously enthusiastic leader, Steve called me one day to say he had an idea. He wanted me to become head of engineering for the online search and advertising business that would later be relaunched as Bing, one of Microsoft's first businesses born in the cloud.

For context, search engines generate revenue through a form of advertising known as an auction. Advertisers bid on search keywords that match their product or service; the winning bid gets an opportunity to display a relevant advertisement on the search results page. Search for a car and a car dealership has likely paid to be displayed prominently on your results page. Delivering that purchase experience both from the consumer and the advertiser perspective is computationally expensive and sophisticated. And while Microsoft was struggling with low market share in search, Steve had invested in it because it would require the company to compete in a sector beyond Windows and Office and build great technology—which he saw as the future of our industry. There was tremendous pressure for Microsoft to answer Amazon's growing cloud business. This was the business he was inviting me to join.

"You should think about it, though," Steve added. "This might be your last job at Microsoft, because if you fail there is no parachute. You may just crash with it." I wondered at the time whether he meant it as a grim bit of humor or as a perfectly straightforward warning. I'm still not quite sure which it was.

Despite the warning, the job sounded intriguing. I was running an emerging new business within Microsoft Dynamics. I had taken over from Doug Burgum who later would become the governor of North Dakota. Doug was an inspirational leader who mentored me to become a more complete leader. He thought about business and work not in isolation but as part of a broader societal fabric and a core part of one's life. Some of the lessons I learned from Doug are today an important part of who I am as a leader. Leading the Dynamics team was a dream job. For the first time, I was getting the chance to run a business end to end. I had spent nearly five years preparing for this job. I had all the relationships, inside and outside Microsoft, to drive the Dynamics business forward. But Steve's offer was essentially pushing me out of my comfort zone. I'd never worked in a consumer-facing business and had not really tracked Microsoft's search engine efforts or our early attempts to build cloud infrastructure. So one night, after a long day at work, I decided to drive over to Building 88, which housed the Internet search engineering team. I wanted to walk the hallways and see who these people were. How else could I empathize with the team I was being asked to lead? It was about 9 p.m., but the parking lot was packed. I'd expected to see a few stragglers finishing up their day but, no, the whole team was there working at their desks and eating take-out food. I didn't really talk to anyone. But what I observed caused me to wonder: What gets people to work like this? Something important must be happening in Building 88.

Seeing the team that night, their commitment and dedication, clinched it for me. I told Steve, "Okay, I'm in." What color was my parachute? I didn't have one.

I was entering a new world, and the move proved to be fortuitous. Little did I know it would be my proving ground for future leadership and the future of the company.

Very quickly I realized we would need four essential skills to build an online, cloud-based business that would be accessed primarily from mobile phones rather than desktop computers.

First, I thought I knew a lot about distributed computing systems, but suddenly I realized I had to completely relearn these systems because of the cloud. A distributed system, simply put, is how software communicates and coordinates across networked computers. Imagine hundreds of thousands of people typing in search queries at the same time. If those queries landed in just one server somewhere in a room on the West Coast, it would break that server. But now imagine those queries being distributed evenly across a network of servers. The vast array of computing power would enable delivery of instant, relevant results to the consumer. And, if there's more traffic, just add more servers. This elasticity is a core attribute of cloud computing architecture.

Second, we had to become great at consumer product design. We knew we needed great technology, but we also understood we needed a great experience, one you want to engage with time and again. Traditional software design mapped out what developers thought a product should look like in a year's time, when it would finally go to market. Modern software design involves online products updated through continuous experimentation. Designers offer Web pages in "flights," so an old version of Bing is delivered to some searchers while an untested new version reaches others. User scorecards determine which is the most effective. Sometimes, seemingly tiny differences can mean a lot. Something as simple as the color or size of a type font may profoundly impact the willingness of consumers to engage, triggering behavioral variations that may be worth tens of millions in revenue. Now Microsoft had to master this new approach to product design.

Third, we had to be great at understanding and building two-sided markets—the economics of a new online business. On one side are the consumers who go online for search results, and on the other side are the advertisers who want their businesses to be found. Both are needed to succeed. This creates the auction effect I was describing earlier. Both sides of the business are equally important, and designing the experience for both sides is crucial. Attracting more and more searchers obviously makes it easier to attract more and more advertisers. And showing the right advertisements is crucial to delivering relevant results. So, "bootstrapping" the online auction and improving search results' relevance would prove to be a vital challenge.

Finally, we needed to be great at applied machine learning (ML). ML is a very rich form of data analytics that is foundational to artificial intelligence. We needed a sophisticated understanding of how to do two things at once—discern the intent of someone searching the Web and then match that intent with accurate knowledge gained from crawling the Web, ingesting and understanding information.

Ultimately, Bing would prove to be a great training ground for building the hyper-scale, cloud-first services that today permeate Microsoft. We weren't just building Bing, we were building the foundational technologies that would fuel Microsoft's future. Building Bing taught us about scale, experimentation-led design, applied ML, and auction-based pricing. These skills are not only mission critical at our company, but highly sought after throughout today's technology universe.

But we started very much behind in search; we had yet to launch a product that could compete with Google. So I hit the road, meeting with executives from Facebook, Amazon, Yahoo, and Apple to evangelize our emerging search engine. I wanted to make deals, but I also wanted to learn more about how they engineered their products to stay fresh. I found that the key was agility, agility, agility. We needed to develop speed, nimbleness, and athleticism to get the consumer experience right, not just once but daily. We needed to set and repeatedly meet short-term goals, shipping code at a more modern, fast-paced cadence.

To accomplish this, we needed to periodically gather all of the decision makers in a war-room setting. In September 2008 I called together the search engineers for the first of these meetings, which we casually called Search Checkpoint #1. (Perhaps we should have been more creative with the name, because it has stuck and now we're at a checkpoint in the many hundreds.) We had decided to launch Bing in June 2009—a new search engine and a new brand. I learned a lot about creating urgency and mobilizing leaders with different skills and backgrounds toward one common goal in what was a new area for Microsoft. I realized that in a successful company it is as important to unlearn some old habits as it is to learn new skills.

My learning during this time was greatly accelerated by the hiring of Dr. Qi Lu as head of all online services at Microsoft. Qi had been an executive at Yahoo and was intensely recruited throughout Silicon Valley. Steve, Harry Shum, today our head of

AI and research, and I had gone down to the Bay Area to spend an afternoon talking to Qi. On the flight back Steve said to me, "We should get him, but if you don't want to work for him, that will be a problem." Having just met with Qi, I knew that he was someone from whom I could learn a lot and Microsoft could benefit. So, I did not hesitate in supporting the hiring of Qi to Microsoft, even though in some sense it was stalling my own promotion. I realized that my own professional growth would come from working for and learning from Qi during my time in our online business. Later Qi would become an important member of my senior leadership team during the first few years I was CEO. Qi eventually left the company, but he continues to be a trusted friend and advisor.

Over time, Yahoo integrated Bing as its search engine, and together we powered a quarter of all U.S. searches. The search engine that many had said should be shuttered in its early days of struggle continued to win an expanding share of the market, and today it is a profitable multi-billion-dollar business for Microsoft. Just as important, though, was how it helped to jump-start our move to the cloud.

As was so often the case at Microsoft, there were other experiments elsewhere in the company aimed at the same problem, leading to internal competition and even fiefdoms. Since 2008, Ray Ozzie had been incubating a highly secretive cloud infrastructure product with the code name Red Dog. A longtime Microsoft reporter, Mary Jo Foley, came across a job advertisement for a Red Dog engineer and wrote a piece speculating that this project must be Microsoft's answer to Amazon's AWS.

At some point during my time at Bing, I met with the Red Dog team to explore how we might work together. I quickly realized that Microsoft's storied server and tools business (STB), where products like Windows Server and SQL Server had been invented and built and where Red Dog was housed, was worlds apart from Bing. STB was Microsoft's third largest group by revenue after Office and Windows. They were the deep distributed systems experts. But when I contrasted STB with Bing a few things were apparent. They lacked the feedback loop that comes from running an at-scale cloud service. I realized that they were caught up in the local maxima of servicing their existing customer base and were not learning fast enough about the new world of cloud services. And the Red Dog team was a side effort that was ignored by the mainstream of the STB leadership and organization.

In late 2010, Ray Ozzie announced in a long internal memo that he was leaving Microsoft. He wrote in his departure email, "The one irrefutable truth is that in any large organization, any transformation that is to 'stick' must come from within." While Red Dog was still in incubation and had booked little revenue, he was correct that the transformation of Microsoft would come from within. Steve had already proclaimed that the company was all-in on the cloud, having invested $8.7 billion in research and development, much of it focused on cloud technologies. But even though engineers were working on cloud-related technologies, a clear vision of a Microsoft cloud platform had not yet surfaced—to say nothing of a real-world revenue stream.

Right around that time, Steve asked that I lead STB, which today has evolved into Microsoft's cloud and enterprise business. I was given this news of my new role not even a week before I got the job. Steve had a sense that we needed to move faster to the cloud. He had personally and aggressively driven the transformation of our Office business to the cloud. He wanted us to be equally bold when it came to cloud infrastructure. When I took over our fledgling cloud business in January 2011, analysts estimated that cloud revenues were already multi-billions of dollars with Amazon in the lead and Microsoft nowhere to be seen. Meanwhile, revenues from our cloud services could be counted in the millions, not the billions. Although Amazon did not report its AWS revenues in those days, they were the clear leader, building a huge business without any real challenge from Microsoft. In his annual letter to shareholders in April 2011, just as I was beginning my new role, Amazon CEO Jeff Bezos gleefully offered a short course on the computer science and economics underlying their burgeoning cloud enterprise. He wrote about Bayesian estimators, machine learning, pattern recognition, and probabilistic decision making. "The advances in data management developed by Amazon engineers have been the starting point for the architectures underneath the cloud storage and data management services offered by Amazon Web Services (AWS)," he wrote. Amazon was leading a revolution and we had not even mustered our troops. Years earlier I had left Sun Microsystems to help Microsoft capture the lead in the enterprise market, and here we were once again far behind.

As a company, we'd been very publicly missing the mobile revolution, but we were not about to miss the cloud. I would miss working with colleagues at Bing, but I was excited to lead what

I sensed would be the biggest transformation of Microsoft in a generation—our journey to the cloud. I had spent three years, from 2008 to 2011, learning the cloud—pressure-testing its infrastructure, operations, and economics—but as a user, not as a provider of the cloud. That experience would enable me to execute with speed in my new role.

But it wouldn't be easy. The server and tools business was at the peak of its commercial success and yet it was missing the future. The organization was deeply divided over the importance of the cloud business. There was constant tension between diverging forces. On the one hand, the division's leaders would say, "Yes, there is this cloud thing," and "Yes, we should incubate it," but, on the other hand, they would quickly shift to warning, "Remember, we've got to focus on our server business." The servers that had made STB a force within Microsoft and the industry, namely Windows Server and SQL Server, were now holding them back, discouraging them from innovating and growing with the times.

Shortly after I took over, the company issued this statement: "Nadella and his team are tasked with leading Microsoft's enterprise transformation into the cloud and providing the technology roadmap and vision for the future of business computing." Steve had said the transformation would not happen overnight, but we were running out of time.

I had a very good idea about where we needed to go, but I realized that my real task was to motivate the pride and desire in the STB leaders to go there with me. Sure, I had a point of view, but I also recognized this was a team that cared deeply about enterprises, those customers with exacting and sophisticated computing needs. I wanted to build on their institutional knowledge and so I set out first to learn from the team I was to lead, and, hopefully, to earn the team's respect. Only then could we go boldly together to a new and better place.

Leadership means making choices and then rallying the team around those choices. One thing I had learned from my dad's experience as a senior Indian government official was that few tasks are more difficult than building a *lasting* institution. The choice of leading through consensus versus fiat is a false one. Any institution-building comes from having a clear vision and culture that works to motivate progress both top-down and bottom-up.

In business school I had read *Young Men and Fire*, a book by Norman Maclean (best known for *A River Runs Through It*). It tells the story of a tragic forest fire that killed thirteen "smoke-jumpers" (parachuting firefighters) in 1949 and the investigation that followed. What I remembered was the lesson that went unheeded: the urgent need to build shared context, trust, and credibility with your team. The lead firefighter, who ultimately escaped the blaze, knew that he had to build a small fire in order to escape the bigger fire. But no one would follow him. He had the skills to get his men out of harm's way, but he hadn't built the shared context needed to make his leadership effective. His team paid the ultimate price.

I was determined not to make the same mistake.

Like that lead firefighter, I had to convince my team to adapt a counterintuitive strategy—to shift focus from the big server and tools business that paid everyone's salary to the tiny cloud business with almost no revenue. To win their support, I needed to build shared context. I decided not to bring my old team from Bing with me. It was important that the transformation come from within, from the core. It's the only way to make change sustainable.

The team I inherited was more like a group of individuals. The poet John Donne wrote, "No man is an island," but he'd feel otherwise had he joined our meetings. Each leader in the group was, in essence, CEO of a self-sustaining business. Each lived and operated in a silo, and most had been doing so for a very long time. My portfolio had no center of gravity, and to make matters worse, many thought they should have gotten my job. Their attitude was one of frustration—they were making all this money and now this little squeaky thing called the cloud came along and they didn't want to bother with it.

To break out of this impasse, I met with everyone on the STB leadership team individually, taking their pulse, asking questions and listening. Together we had to see that our North Star would be a cloud-first strategy. Our products and technologies would optimize for the cloud, not just for private servers that resided on an organization's own premises. Though we would be cloud-first, our server strength would enable us to differentiate ourselves as the company that delivered a hybrid solution to customers who wanted both private, on-premise servers and access to the public cloud.

This new framework helped reshape the argument, breaking down the resistance to going all-in on the cloud. I began to notice a new openness to innovation and a search for creative ways to meet the needs of our commercial customers.

Unfortunately, Red Dog, which had become Windows Azure, was still struggling. They were trying to leapfrog with a new approach to cloud computing, but the market was clearly giving them feedback that they first needed to meet their current needs. Mark Russinovich, who was an early member of the Red Dog team and the current CTO of Azure, had a clear road map in mind to evolve Azure. We needed to infuse more resources into the team to execute on that road map.

It was time to move Azure into the mainstream of STB rather than have it be a side project. People, the human element of any enterprise, are ultimately the greatest asset, and so I set about assembling the right team, starting with Scott Guthrie, a very accomplished Microsoft engineer. He had spearheaded a number of successful company technologies focused on developers. I tapped him to lead engineering for Azure on its way to becoming Microsoft's cloud platform—our answer to Amazon Web Services.

Over time, many others from both inside and outside the company joined our effort. Jason Zander, another key leader who built .Net and Visual Studio, joined to lead the core Azure infrastructure. We recruited the highly regarded Big Data researcher Raghu Ramakrishnan from Yahoo and James Phillips who had cofounded the database company Couchbase. We relied heavily on the expertise of Joy Chik and Brad Anderson to advance our device management solutions for the mobile world. Under their leadership we made our first major steps in providing business customers the technology they need to secure and manage Windows, iOS, and Android devices. Julia Liuson took over our Visual Studio developer tools, evolving it to be the tool of choice for any developer regardless of platform or app.

Complementing these world-class engineers was world-class business planning and modeling. Takeshi Numoto moved from the Office team to join STB. Takeshi had been an important member of the team that had strategized and executed the transformation of Office products to a cloud-based, subscription model. And in his role as business lead for STB, he set about building the new commercial model that was based on creating meters to measure consumption of cloud services and inventing new ways to package our products for customers.

One of the early decisions I made was to differentiate Azure with our data and AI capabilities. Raghu and team designed and built the data platform that could help store and process exabyte-scale data. Microsoft was developing machine learning and AI capability as part of our products such as Bing, Xbox Kinect, and Skype Translator. I wanted us to make this capability available to third-party developers as part of Azure.

A key hire for Azure was Joseph Sirosh, who I recruited from Amazon. Joseph had been passionately working in ML for all his professional career, and he brought that passion to his new role at Microsoft. Now our cloud not only could store and compute massive amounts of data, it could also analyze and learn from the data.

The practical value of ML is immense and incredibly varied. Take a Microsoft customer like ThyssenKrupp, a manufacturer in the elevator and escalator business. Using Azure and Azure ML, they can now predict in advance when an elevator or escalator will need maintenance, virtually eliminating outages and creating new value for its customers. Similarly, an insurer like MetLife can spin up our cloud with ML overnight to run enormous actuarial tables and have answers to its most crucial financial questions in the morning, making it possible for the company to adapt quickly to dramatic shifts in the insurance landscape—an unexpected flu epidemic, a more-violent-than-normal hurricane season.

Whether you are in Ethiopia or Evanston, Ohio, or if you hold a doctorate in data science or not, everyone should have that capability to learn from the data. With Azure, Microsoft would democratize machine learning just as it had done with personal computing back in the 1980s.

To me, meeting with customers and learning from both their articulated and unarticulated needs is key to any product innovation agenda. In my meetings with customers I would usually bring other leaders and engineers along so that we could learn together. On one trip to the Bay Area, we met with several start-ups. It became clear that we needed to support the Linux operating system, and we had already taken some rudimentary steps toward that with Azure. But as Scott Guthrie and our team walked out of those meetings that day, it was certain that we needed to make first-class support for Linux in Azure. We made that decision by the time we got to the parking lot.

This may sound like a purely technical dilemma, but it also posed a profound cultural challenge. Dogma at Microsoft had long held that the open-source software from Linux was the enemy. We couldn't afford to cling to that attitude any longer. We had to meet the customers where they were and, more importantly, we needed to ensure that we viewed our opportunity not through a rearview mirror, but with a more future-oriented perspective. We changed the name of the product from Windows Azure to Microsoft Azure to make it clear that our cloud was not just about Windows.

To scale our cloud business, we not only needed to build the right technology but we needed to run a service to meet the exacting needs of some of the world's largest customers. We were already running at-scale services such as Bing, Office 365, and Xbox Live. But with Azure we were now powering thousands of other businesses every minute of every day.

Our team had to learn to embrace what I called "live site first" culture. The operational culture was as important as any key technology breakthrough. We would have on a single Skype call dozens of engineers plus our customer-facing field teams, all of whom would swarm together to coordinate and fix any problem. And every such incident would lead to rigorous root-cause analysis so that we could continuously learn and improve. I would, from time to time, join these calls to see our engineers in action. The key is to not have the top leaders infuse fear or panic but to help foster the actions that fix the issue at hand and the learning from it.

Today Microsoft is on course to have its own $20 billion cloud business. We had looked beyond the packaged products that had made Microsoft one of the most valuable companies in the world to see greater opportunity in our cloud platform, Azure, and cloud services like O365, the online version of our hugely popular Office productivity suite. We are investing in and improving these new products, building new muscle as a service provider, and embracing Linux and other open-source efforts, all while keeping focus on our customers.

The cloud business taught me a series of lessons I would carry with me for years to come. Perhaps the most important is this: A leader must see the external opportunities and the internal capability and culture—and all of the connections among them—and respond to them before they become obvious parts of the conventional wisdom. It's an art form, not a science. And a leader will not always get it right. But the batting average for how well a leader does this is going to define his or her longevity in business. It's an insight that would serve me well when an even bigger set of challenges was presented to me as CEO.

CHAPTER 3

New Mission, New Momentum

Rediscovering the Soul of Microsoft

On the morning of February 4, 2014, the day I would be formally introduced as CEO for the first time, I drove to campus early, rehearsing the messages I wanted to land with employees on day one. Over the Thanksgiving break I had banged out a ten-page memo responding to several questions the board of directors had posed during the search process. I then refined it on a road trip that took me to Dell in Texas and Hewlett Packard in Silicon Valley. The questions required a lot of soul-searching. What is my vision? What is the strategy to achieve it? What does success look like and where to get started? Now, a few months later, I reflected on what I had written and the process that led to this day.

Finding the next CEO had been a long journey. Steve had surprised everyone by announcing in August that he would retire, just after he had led a major reorganization of the company and on the eve of announcing the $7.2 billion deal with the Finnish smartphone manufacturer Nokia. Throughout the fall, reporters routinely speculated about who would be named as his replacement. Would it be an outsider like Ford Motor Company's CEO, Alan Mulally? Or the executive of a Microsoft acquisition like Tony Bates of Skype or Stephen Elop of Nokia? Several of us were asked to put our ideas on paper for the board of directors as a sort of audition for the job.

In my memo to the board, I'd drawn on more than twenty years of experience inside the company, but I'd also drawn on something Steve Ballmer, the departing CEO, told me. He encouraged me to be my own man. In other words, don't try to please Bill Gates or anyone else. "Be bold, be right," he told me. Bill and Paul Allen founded Microsoft. Bill and Steve built Microsoft. As founder, Bill had famously recruited Steve out of Stanford business school in 1980 to become his first business manager. Steve, the passionate leader, salesman, and marketer, and Bill the technology visionary whose voracious reading and "think weeks" kept Microsoft ahead of envious competitors— together they formed one of the most iconic business partnerships in history, a pioneer in computing that would make Microsoft the most valuable company on the planet. Not only did they build great products, but they shaped hundreds of executives who today run global businesses everywhere, including me. They had given me more and more responsibility over the years, and taught me that our software can impact not just the lives of computer hobbyists but entire societies and economies.

Despite my devotion to what they had built, Steve didn't want me to be confused. He was inviting me to throw dogma out the window. He knew more than anyone that the company had to change, and he selflessly stepped out of his role as CEO to ensure the change happened in a deep way. As a consummate insider, I was being told to start anew, to refresh the browser and load a new page—the next page in Microsoft's history. And so, my memo to the board called for a "renewal of Microsoft." It would require embracing more ubiquitous computing and ambient intelligence. This means humans will interact with experiences that span a multitude of devices and senses. All these experiences will be powered by intelligence in the cloud and also at the edge where data is being generated and interactions with people are taking place. But this renewal would only happen, I wrote, if we prioritized the organization's culture and built confidence both inside and outside the company. It would be only too easy to continue to live off our past successes. We had been like kings, albeit now in a threatened kingdom. There were ways to cash-cow this business and drive short-term return, but I believed we could build long-term value by being true to our identity and innovating.

I pulled into a parking space outside Studio D, home of our Xbox development team. This part of the campus didn't even exist when I started at Microsoft in 1992, and yet now it was surrounded by several dozen low-slung office buildings as far as the eye could see. Over the next hour Studio D's glassy three-story atrium would fill with employees invited to an all-company meeting and webcast. I knew they were hopeful but also skeptical. One had only to look at a few industry charts to see why. After decades of steady growth in worldwide PC shipments, sales had peaked and were now in decline. Quarterly PC shipments were now around 70 million, while smartphone shipments were reaching over 350 million. This was bad news for Microsoft. Every PC sold meant a royalty payment to Microsoft. To make matters worse, not only were PC sales soft, but so was interest in Windows 8, launched eighteen months earlier. Meanwhile Android and Apple operating systems were surging, reflections of the smartphone explosion that we at Microsoft had failed to lead and barely managed to participate in. As a result, Microsoft's stock, long a blue chip investment, had been treading water for years.

Internally, the story was equally dire. That year, our annual employee poll revealed that most employees didn't think we were headed in the right direction and questioned our ability to innovate. To bring this home, Jill Tracie Nichols, my chief of staff at the time, shared new focus group feedback from hundreds of employees so I could get a real-time pulse of the organization in the midst of the change. The company was sick. Employees were tired. They were frustrated. They were fed up with losing and falling behind despite their grand plans and great ideas. They came to Microsoft with big dreams, but it felt like all they really did was deal with upper management, execute taxing processes, and bicker in meetings. They believed that only an outsider could get the company back on track. None of the names on the rumored list of internal CEO candidates resonated with them—including mine.

In an intense prep session two days before the announcement Jill and I sparred on how to inspire this disheartened group of brilliant people. In some ways, I was annoyed by what felt like lack of accountability and finger pointing. She stopped me mid-riff with "You're missing it, they are actually hungry to do more, but things keep getting in their way." Job one was to build hope. This was day one of our transformation—I knew it must start from within.

A few minutes later, I stood onstage for a photo that would soon go viral. It captured the smiling faces of Bill Gates, Steve Ballmer, and me, the only CEOs in Microsoft's-forty-year history. The image I remember even more vividly, however, is looking into the eyes of hundreds of Microsoft employees in the audience waiting for my presentation, their faces reflecting hope, excitement, and energy mingled with anxiety and a touch of frustration. Like me, they'd come to Microsoft to change the world, but now were frustrated by our company's stalled growth. They were being wooed by competitors. Saddest of all, many felt the company was losing its soul.

Steve kicked things off with a moving and encouraging speech. Bill spoke next, his dry sense of humor immediately present. Surveying the room, he feigned surprise at what a large market share Windows Phone enjoyed in this room. Then he got down to business. Bill succinctly captured the challenge and the opportunity that lay ahead. "Microsoft was founded based on a belief in the magic of software, and I'd say that opportunity today is stronger than it's ever been. The magic of what we can do for people at work and at home with our software is totally in front of us. We've got some amazing strengths with the Windows platform, the things we're doing in the cloud, with Office. And we've got some challenges. There are a lot of people out there on the cloud doing interesting things. There's a lot of mobile activity, which we've got a slice of, but not as big a slice as we need to have." Then he called me forward.

When the applause subsided, I wasted no time in calling my colleagues and teammates to action. "Our industry does not respect tradition. What it respects is innovation. It's our collective challenge to make Microsoft thrive in a mobile-first and a cloud-first world." If there was any one theme I wanted to emphasize that day, it was that we must discover what would be lost in the world if Microsoft just disappeared. We had to answer for ourselves, what is the company about? Why do we exist? I told them it was time for us to rediscover our soul—what makes us unique.

One of my favorite books is Tracy Kidder's *The Soul of a New Machine* about another tech company, Data General, in the 1970s. In it, Kidder teaches us that technology is nothing more than the collective soul of those who build it. The technology is fascinating, but even more fascinating is the profound obsession of its designers. And so what is soul in this context of a company? I don't mean soul in a religious sense. It is the thing that comes most naturally. It is the inner voice. It's what motivates and provides inner direction to apply your capability. What is the unique sensibility that we as a company have? For Microsoft that soul is about empowering people, and not just individuals, but also the institutions they build—enterprises like schools, hospitals, businesses, government agencies, and nonprofits.

Steve Jobs understood what the soul of a company is. He once said that "design is the fundamental soul of a man-made creation that ends up expressing itself in successive outer layers of the product or service." I agree. Apple will always remain true to its soul as long as its inner voice, its motivation, is about great design for consumer products. The soul of our company is different. I knew that Microsoft needed to regain its soul as the company that makes powerful technology accessible to everyone and every organization—democratizing technology. When I first wore HoloLens, Microsoft's holographic computer, I thought about how it might be used by large enterprises for design and in schools and hospitals, not just how much fun playing Minecraft will be.

It's not that we had lost our soul, but we needed renewal, a renaissance. In the 1970s, Bill and Paul Allen started Microsoft with the goal of helping to put a computer on every desk and in every home. That was a bold, inspiring, and audacious ambition, one they accomplished. Democratizing and personalizing technology. How many organizations can say they achieved their founding mission? There is no way I would become CEO of a Fortune 500 company if it was not for the democratizing force of Microsoft all over the world. But the world had changed, and it was time for us to change our view of the world.

Worldview is an interesting term, rooted in cognitive philosophy. Simply put, it is how a person comprehensively sees the world—across political, social, and economic borders. What are the common experiences we all share? The question I had been asking before becoming CEO—why do we exist?—forced me to change my tech worldview, and, similarly, now every leader at Microsoft was changing theirs as well. We no longer lived in a PC-centric world. Computing was becoming more ubiquitous. Intelligence was becoming more ambient, meaning computers could observe, collect data, and turn that feedback into insights. We were seeing an ever-increasing wave of digitization of our life, business, and our world more broadly. This was made possible by an ever-growing network of connected devices, incredible computing capacity from the cloud, insights from big data, and intelligence from machine learning. I simplified all of this and encouraged Microsoft to become "mobile-first and cloud-first." Not PC-first or even Phone-first. We needed to envision a world where the mobility of the human experience across all devices was what mattered, and the cloud made that mobility possible, enabling the new generation of intelligent experiences. The transformation we would undertake across all parts of our businesses would help Microsoft and our customers thrive in this new world.

It might be easy to be motivated to change through envy. We could envy what Apple had built with its iPhone and its iPad franchise, or what Google had created with its low-cost Android phones and tablets. But envy is negative and outer-directed, not driven from within, and so I knew that it wouldn't carry us very far down the path to true renewal.

We could also motivate ourselves through competitive zeal. Microsoft is known for rallying the troops with competitive fire.

The press loves that, but it's not me. My approach is to lead with a sense of purpose and pride in what we do, not envy or combativeness.

Our senior leadership team recognized a gap in the competitive landscape that Microsoft was in the unique position to fill. You see, while our competitors defined their products as mobile, we could be about the mobility of human experiences, experiences made possible by our cloud technologies. These two trends together, mobile and cloud, were fundamental to our transformation. In fact, our marketing chief, Chris Capossela, would produce an ad for the Microsoft Cloud based on a speech I gave on this subject. In the ad, Spain's Real Madrid soccer team is seen attacking, nimbly sprinting toward the goal as Grammy Award–winning hip hop artist Common tells the audience, "We live in a world of mobile technology, but it is not the device that is mobile. It is you."

The ad, though made for our broader audiences, also helped remind us of our core, the soul that we almost lost. Microsoft had led the PC Revolution by enabling the highest-volume, most-affordable computing devices. But Google, with its free Android operating system, found a way to undercut Windows, something we did not react to quickly enough. In 2008, the Linux-based Android smartphone gobbled up market share and today it runs on more than one billion activated devices.

Looking back, the Nokia deal announced in September 2013, five months before I would become CEO, became another painful example of this loss. We were desperate to catch up after missing the rise of mobile technology. Nokia, which had overtaken Motorola in the 1990s as the largest mobile phone manufacturer, had lost ground to Apple's iPhone and Google's Android phones. Nokia fell from the market-share leader in mobile to number three. By 2012, in a risky move to regain ground, Nokia's CEO Stephen Elop announced it would make Windows the primary operating system for Nokia smartphones. Nokia and Microsoft did make progress (achieving double-digit market share in some European countries), but we still remained a distant third. The hope behind the acquisition was that combining the engineering and design teams at Nokia with software development at Microsoft would accelerate our growth with Windows Phone and strengthen our overall device ecosystem. The merger could be the big, dramatic move Windows needed to catch up with iOS and Android in mobile.

The press criticized the idea and the Microsoft board was resistant. Over the summer, while still in negotiations to buy Nokia outright, Steve Ballmer asked the members of his leadership team, his direct reports, to vote thumbs-up or thumbs-down on the deal. He wanted a public vote to see where the team was on the matter. I voted no. While I respected Steve and understood the logic of growing our market share to build a credible third ecosystem, I did not get why the world needed the third ecosystem in phones, unless we changed the rules.

A few months after I became CEO, the Nokia deal closed, and our teams worked hard to relaunch Windows Phone with new devices and a new operating system that came with new experiences. But it was too late to regain the ground we had lost. We were chasing our competitors' taillights. Months later, I would have to announce a total write-off of the acquisition as well as plans to eliminate nearly eighteen thousand jobs, the majority of them because of the Nokia devices and services acquisition. It was heartbreaking to know that so many talented people who gave so much of themselves to their work would lose their jobs.

There are many lessons a leader can take from the Nokia acquisition. Buying a company with weak market share is always risky. What we needed most was a fresh and distinctive approach to mobile computing. Where we went wrong initially was failing to recognize that our greatest strengths were already part of the soul of our company—inventing new hardware for Windows, making computing more personal, and making our cloud services work across any device and any platform. We should only be in the phone business when we have something that is really differentiated.

We did ultimately follow up on this key insight. We focused our efforts more narrowly by building Windows Phone with organizations in mind. For example, these business customers now love Continuum, a feature that makes it possible for a phone to replace a PC. To further participate in mobile, we also made Office run across devices. In retrospect, what I regret most is the impact these layoffs had on very talented, passionate people in our phone division.

———

Early in my new role, Bill Gates and I walked together from one building to the next to meet with a reporter from *Vanity Fair*.

Bill had decided to remain on the board, but would step down as chairman. The foundation he'd cofounded with his wife, Melinda, would now be his primary focus, but he remained passionate about software and about Microsoft. On our walk, he enthusiastically talked about a new product that would blur the lines between a document and a website. We brainstormed how to develop an architecture that would enable rich capabilities for composing a report, but instead of a static page it would have all the richness of an interactive website. We quickly got into the weeds, volleying ideas back and forth about visualization data structures and storage systems. At one point Bill looked at me, smiled, and said it was good to be talking software engineering.

I knew that part of rediscovering the company's soul was to bring Bill back, to engage him more deeply in the technical vision for our products and services. Reviewing a software product with Bill is the stuff of legend at Microsoft. In his comical 1994 novel, *Microserfs*, Douglas Coupland wrote a humorous sketch of Bill's influence on a Microsoft programmer. A developer named Michael locks himself in his office at 11 a.m. after getting a flame mail from Bill who had reviewed some of his code. No one on his floor had ever been flamed by Bill personally. "The episode was tinged with glamour and we were somewhat jealous." By 2:30 a.m., concerned about Michael, his team goes to the twenty-four-hour Safeway for "flat" food that can be slipped underneath his door. While this exaggerated legend doesn't exactly capture the culture I hoped to create, I knew that getting our founder back in the product-development loop would inspire our people and up our game.

Over the first several months of my tenure, I devoted a lot of time to listening, to anyone and everyone just as I had promised to do in that Thanksgiving memo to the board. I met with all of our leaders and made a point of going out as I always had to meet with partners and customers. As I listened, there were two questions I was still trying to answer. The first, why are we here? Answering this question would be central to defining the company for years to come. The second question was, what do we do next? There is that great closing scene in *The Candidate* when Robert Redford, having finally won the election, pulls his advisor into a room and asks, "What do we do now?" For starters, I decided to listen.

Straightaway I heard from hundreds of employees at every level and in every part of the company. We held focus groups to allow people to share their opinions anonymously as well. Listening was the most important thing I accomplished each day because it would build the foundation of my leadership for years to come. To my first question, why does Microsoft exist, the message was loud and clear. We exist to build products that empower others. That is the meaning we're all looking to infuse into our work. I heard other things as well. Employees wanted a CEO who would make crucial changes, but one who also respected the original ideals of Microsoft, which had always been to change the world. They wanted a clear, tangible and inspiring vision. They wanted to hear more frequently about progress in transparent and simple ways. Engineers wanted to lead again, not follow. They wanted to up the coolness. We had technology the press would fawn over in Silicon Valley, such as leading-edge artificial intelligence, but we weren't showing it off. What they really demanded was a road map to remove paralysis. For example, Google made headlines with glitzy demonstrations of their artificial intelligence experiments while we had world-class speech and vision recognition and advanced machine learning that we kept under wraps. The real challenge I was contemplating, though, was how do we take our technologies and do things that speak to our identity and add unique value for our customers?

On my second question, where do we go from here, I became convinced that the new CEO of Microsoft needed to do several things very well right away, during the first year.

- Communicate clearly and regularly our sense of mission, world-view, and business and innovation ambitions.
- Drive cultural change from top to bottom, and get the right team in the right place.
- Build new and surprising partnerships in which we can grow the pie and delight customers.
- Be ready to catch the next wave of innovation and platform shifts. Reframe our opportunity for a mobile- and cloud-first world, and drive our execution with urgency.
- Stand for timeless values, and restore productivity and economic growth for everyone.

This list does not suggest a formula for success since even today Microsoft is still very much in the midst of change. We will not know the lasting impact of our approach for some time.

But between the summers of 2014 and 2015, we pushed for change with a steady drumbeat. Having listened with great intensity and curiosity during those first several months, it was time to act and to do so with confidence and conviction. My first title at Microsoft had been "evangelist," a common term in technology for someone who drives a standard or product to achieve critical mass. Now here I was evangelizing the notion that we needed to rediscover our soul. The mission of a company is in many ways a statement about its soul, and that's where I went first.

To make things real and drive fidelity of the ideas through an organization of 100,000-plus people operating across more than 190 countries we developed a clear connection between our mission and our culture. We defined our mission, worldview, ambitions, and culture in one page—no small feat for a company that loves massive PowerPoint decks. That was the relatively easy part. The harder part was to not tweak it—to let it stand. I'd want to edit a word here or there, add a row, just tinker with it before each speech. Then, I'd be reminded again "consistency is better than perfection."

In the years prior to taking on my new role, the executive team spent far too much time trying to explain the massive company and our strategy. We needed a shared understanding. The simple framework we came up with catalyzed people to bring these ideas to life.

The work in these first few years of my tenure was all about getting the flywheel of change spinning. Sure, it took regular communications, but it also took discipline and consistency on my part and that of the senior leadership team. We needed to inspire and drive change. We challenged ourselves, "At the end of the next year if we were tried in a court of law and the charge was that we failed to pursue our mission, would there be enough evidence to convict us?" Just saying interesting things wasn't enough. I, all of us, had to do them. And our employees had to see how everything we did reinforced our mission, ambitions, and culture. And then they needed to start doing the same.

Our three ambitions defined how we organized teams and reported results. Our mission guided where I visited and who I met while I was there. My travel itinerary frequently started with a visit to a school or hospital in the community. I particularly enjoyed ceremonies with the indigenous peoples in Colombia and New Zealand, learning about how they used Microsoft technology to preserve their history and traditions for generations and how they think about growth. Beyond this, we were greenlighting mothballed products and projects, inviting new partnerships with competitors, showing up in surprising places, making accessibility a first-class citizen in our product design efforts, and constantly traveling the world to engage our people, partners, and customers.

On Thursday, July 10, 2014, only a few days into the start of Microsoft's new fiscal year, I sent an all-company email, a sort of manifesto, at 6:02 a.m. so that it landed in in-boxes at the beginning of the day in all U.S. time zones and before the weekend for employees around the world. We are a global company and we needed to think like one. "In order to accelerate our innovation, we must rediscover our soul—our unique core. We must all understand and embrace what only Microsoft can contribute to the world and how we can once again change the world. I consider the job before us to be bolder and more ambitious than anything we have ever done. Microsoft is the productivity and platform company for the mobile-first, cloud-first world. We will reinvent productivity to empower every person and every organization on the planet to do more and achieve more."

I wrote that productivity for us goes well beyond documents, spreadsheets, and slides. We will obsess over helping people who are swimming in a growing sea of devices, apps, data, and social networks. We will build software to be more predictive, personal, and helpful. We will think about customers as "dual users," people who use technology for their work, their school, and their personal digital life. In the email I inserted the image of a target and in its center appeared the words, "digital work and life experiences," surrounded by our cloud platform and computer devices. Soon there will be 3 billion people connected to the Internet, sensors, and the Internet of Things (IoT). Yes, PC sales were slowing, and so we needed to convert Nietzsche's "courage in the face of reality" into "courage in the face of opportunity." We needed to win the billions of connected devices, not fret about a shrinking market.

Employees responded immediately. In just the first twenty-four hours I heard from hundreds of employees in every part of the company and in every part of the world. They said the language of empowering everyone on the planet to achieve more inspired them personally, and they saw how it applied to their daily work, whether they were a coder, designer, marketer, or customer-support technician. Many offered helpful suggestions and ideas. One of my favorites was to challenge conventional thinking more. Why is Xbox a box since traditional television and cable boxes are fading? What if Kinect, our motion-sensing technology used for video games and robotics, came with wings or wheels so it could go fetch lost keys or wallets? Many wrote to me to say that after years of frustration they felt a new energy. I was determined not to squander that.

The press covering Microsoft were given copies of the email, and immediately offered their takes on Microsoft's future under my leadership. *The New York Times* focused on the cultural change under way. *The Washington Post* delighted in "my knack for squeezing literary references into the spaces between his required lines." *Bloomberg* warned that if Microsoft was going to succeed in both the corporate and consumer world with productivity as its focus, "it will need to deliver some products that live up to the rhetoric." They were right. We wanted customers not just to use our products, but to love them.

Articulating our core *raison d'être* and business ambitions was a good first step. But I also needed to get the right people on the bus to join me in leading these changes. A few weeks later I announced that Peggy Johnson, a longtime Qualcomm executive, would join as head of business development, striking deals to acquire and partner with exciting new products and services. Within a few weeks we bought Minecraft, the popular online game, which we knew would boost engagement with our cloud and our devices. A few weeks after that I announced that Kathleen Hogan, who was leading our global consulting and support business, with experience at McKinsey, a worldwide management consulting firm, and Oracle, would become our Chief People Officer and my partner in the cultural transformation to come. I persuaded Kurt Delbene, who had once led our Office division, to return as Chief Strategy Officer after he was handpicked by President Obama to fix Healthcare.org, the nation's health insurance website. We had two people overseeing marketing, and I chose to give the entire role to Chris Capossela. Scott Guthrie, who had been my engineering partner in building the cloud business, was chosen to lead Cloud and Enterprise, our fastest growing business.

Over time these changes meant that some executives left. They were all talented people, but the senior leadership team needed to become a cohesive team that shared a common worldview. For anything monumental to happen—great software, innovative hardware, or even a sustainable institution—there needs to be one great mind or a set of agreeing minds. I don't mean yes-men and yes-women. Debate and argument are essential. Improving upon each other's ideas is crucial. I wanted people to speak up. "Oh, here's a customer segmentation study I've done." "Here's a pricing approach that contradicts this idea." It's great to have a good old-fashioned college debate. But there also has to be high quality agreement. We needed a senior leadership team (SLT) that would lean into each other's problems, promote dialogue, and be effective. We needed everyone to view the SLT as his or her *first* team, not just another meeting they attended. We needed to be aligned on mission, strategy, and culture.

I like to think of the SLT as a sort of Legion of Superheroes, with each leader coming to the table with a unique superpower to contribute for the common good. Amy is our conscience, keeping us intellectually honest and accountable for doing what we committed to do. Kurt pushes us on being rigorous about our strategy and operations. Product leaders like Terry, Scott, Harry, and more recently Rajesh Jha and Kevin Scott push for alignment on product plans, knowing that when we are an inch apart on strategy at the leadership level, our product teams end up miles apart in execution. Brad helps us navigate the ever-evolving legal and policy landscape, always finding just the right position on important global and domestic issues. Kathleen constantly channels the voice of our employees. Peggy does the same for partners, and Chris, Jean-Philippe Courtois, and Judson Althoff for our customers. They are the true heroes of our continuing transformation.

One thing we were all clear on is that beyond the SLT, we needed a broader set of leaders who could be brought into modeling the mission and building the culture we needed. For as long as I could remember, each year the top 150 or so executives would gather for an annual retreat. We left our offices to drive to a remote, mountainous area about two hours from our headquarters. There, we would take up residence at a quiet, comfortable hotel where we would work to get on the same page strategically. This retreat has always been a good idea. Each team shares product plans and performs demos of their latest technology breakthroughs long before the world will experience them. And everyone appreciates having time to reconnect and see their colleagues over meals by the fire. But one aspect of the offsite really bugged me. Here we were with all this talent, all this bandwidth, and all this IQ in one place just talking *at* each other in the deep woods. And frankly, it seemed like most of the talking was about poking holes in each other's ideas. Enough. I figured it was time to hit refresh and experiment. That year, we did several things to symbolize change and to get the top leaders fully on board. I needed them to buy into where we were going, and I needed them to help get us there.

The first change to the retreat was inviting founders of companies we had acquired in the year prior. These new Microsoft leaders were mission-oriented, innovative, born in the mobile-first and cloud-first world. I knew we could learn from their fresh, outside perspective. The only problem was that most of these leaders did not officially "qualify" to go to executive retreats given the person's level in the organization. To make matters worse, neither did the manager, or even their manager's manager. Remember, the retreat had been only for the most senior leaders. Inviting them was not one of my more popular decisions. But they showed up bright eyed, completely ignorant of the history they were breaking. They asked questions. They shared their own journeys. They pushed us to be better.

Another decision, not universally loved, was scheduling customer visits during the retreat. There was more than a little eyerolling and groaning. Why do we have to meet with customers during a retreat? We already meet with them in the normal course of business. Do you think we don't know what our customers really need? But we pushed through the cynicism and met in a conference room the first morning of the retreat. We split up into a dozen or so teams and boarded vans. Each van had a nervous account manager hosting the trip along with a cross section of the company's most senior researchers, engineers, sales, marketing, finance, HR, and operations people, all new to working with one another. The vans headed off in different directions across the Puget Sound region to meet with our customers—schools, universities, large enterprises, nonprofits, startups, hospitals, small businesses, and the like. The executives listened. They learned together. They made new connections with one another. They put down those proverbial guns and discovered new ways Microsoft could fulfill its mission in the world. They experienced the power of having a diverse, cross-company team solving customer's problems together.

Perhaps the most important thing we did during the experimental retreat was engage the leaders in more open and honest dialogue about our cultural evolution. Kathleen Hogan, our Chief People Officer and my partner in this endeavor, knew we needed to get this group's feedback and buy-in. So after a long day of visiting customers in the Seattle area and driving back into the mountains, people were again divided into seventeen random groups of about ten each. They were then pointed to dinner tables with the assignment to share their own account of where the company's culture stood and their ideas on how to evolve it. Some of us guessed this exercise would be futile—a cute nod to leader engagement. We figured these leaders would be tired. They'd be persnickety. They'd want to congregate back with their own friends. They'd say culture was my job or HR's job.

We were dead wrong. Discussions went long into the evening

as the broader executive team gained a common understanding of what others experienced in leading their own teams, and brainstormed ways to create the culture we all aspired to have.

The next morning, Kathleen and each table leader joined me for breakfast to report about what they learned and to share big ideas, ideas born from the previous night's brainstorms. They were passionate, eager to help, and the energy was infectious. In the end, I left the retreat inspired by those ideas, but more importantly I left inspired by the deep engagement and commitment I saw from these leaders. We knew we needed to build on this momentum so we enrolled each of the table discussion leaders in a sort of culture cabinet—a group of trusted advisors and senior leaders who were committed to helping shape and lead the culture change in every part of our company. The change was coming from within.

By the summer of 2015 our leadership team was really coming together and the company was beginning to see momentum. Windows 10, which would be our most ambitious version ever, was nearing launch. The launch of Surface Pro 3 would prove that consumers and businesses alike wanted a tablet that could replace their laptop. We delivered Office for all devices, including the iPhone, and our cloud-based O365 added nearly 10 million subscribers. Azure, Microsoft's cloud platform that competes with Amazon, was growing rapidly. In the months that followed my email to all employees, our leadership team had refined the thinking in that memo and had decided we would affirmatively change the company's mission statement. Our transformation was under way, though we still had a long way to go.

Shortly after the retreat I was scheduled to head for a week-long tour across Asia starting with an important conference in China. Every weekend I would call my mom and talk to her. Since I was traveling that Saturday, I decided to call her before getting on the plane. It was *Ugadi*, New Year's Day for our region of India. I had not realized that, so my mom reminded me and wished me a happy New Year. It was a brief call since I was late to get to the airport, and we talked briefly about the week and all that was going on. We ended the call as usual with her asking me if I was happy with what I was doing and me assuring her that I was. What a blessing because two hours before landing I received a worrisome email from Anu back home asking if I had landed. I sensed something was wrong and after some back-and-forth learned that my mother had unexpectedly passed away. Deeply shaken, I canceled my trip and hurried on to Hyderabad. Over time I would realize that while the death of a parent is painful, my mom is always there in my consciousness. She will always be there. Her calm and mindfulness continue to shape my relationships with people and the world around me to this day.

During that season, I reflected on her role in my life and her constant push to find a sense of contentment and meaning in all I did. This idea was sitting with me through the spring as I prepared to share our new mission and culture with employees globally. That July, I boarded another plane for Orlando, Florida, with a renewed sense of optimism. Every year in July some fifteen thousand customer-facing Microsoft employees gather for a global summit to hear the latest strategies and initiatives and to see demos of new tech products in development. The gathering would be my opportunity to update employees on our progress and enroll them in the changes under way.

With the energy of thousands of colleagues pulsing through the auditorium, I stood backstage rehearsing how I would present our new mission and the imperative to transform our culture. Microsoft people are infamous for presenting a ton of PowerPoint slides when they speak, but I don't like to rely very much on slides or notes. So I was free to just channel what I was thinking and feeling, to let it flow. A computer on every desk and in every home, which Bill and Paul had introduced forty years earlier as the company's mission, was actually more of a goal—an inspiring one for its era. The more I thought about it, the more I questioned what it was that had motivated us to create personal computers in the first place. What was the spirit behind the first line of code ever written for the BASIC interpreter on that primitive computer, the Altair? It was to empower people. And that was still what motivated all of our efforts: *to empower every person and every organization on the planet to achieve more.* We are in the empowerment business, I said as I took the stage, and not just to empower startups and tech-savvy users on the American West Coast, but everyone on the planet. Helping people and their organizations *achieve more* is our sweet spot. That's what informs our decisions and inspires our passion; it's also what makes us different from other companies. We make things that help other people make things and make things happen.

That's the essence of our mission, but our employees and our business partners, ranging from Accenture to Best Buy, Hewlett Packard to Dell, wanted to hear more. They wanted to know our business priorities. To deliver on this promise of empowerment, I said that we must galvanize all of our resources around three interconnected ambitions.

First, we must reinvent productivity and business processes. We needed to evolve beyond simply building individual productivity tools and start designing an intelligent fabric for computing based on four principles—collaboration, mobility, intelligence, and trust. People still do important work as individuals, but collaboration is the new norm, so we build our tools to empower teams. We would aspire to help everyone be productive no matter where they are, regardless of the device they use. Data, apps, and settings—all content—needed to roam across computing experiences. Intelligence is an amazing force multiplier. To be successful amid the explosion of data, people need analytics, services, and agents that use intelligence to help them manage their scarcest resource—time. Finally, trust is the foundation upon which everything we do is built. That's why we've invested heavily in security and compliance that set the standard for enterprises.

Second, we will build the intelligent cloud platform, an ambition closely linked with the first ambition. Every organization today needs new cloud-based infrastructure and applications that can convert vast amounts of data into predictive and analytical power through the use of advanced analytics, machine learning, and AI. From an infrastructure perspective, we would come to deliver on the promise of a global, hyper-scale cloud platform with dozens of unique data centers around the world. We would, over the years, invest billions of dollars each year to build out more and more infrastructure so that our customers could scale their solutions without worrying about their cloud platform's capacity or the complex demands of transparency, reliability, security, privacy, and compliance. Our cloud would be open and offer choice so that we could support a wide range of application platforms and developer tools. We would build our server products to become the edge of our cloud, supporting true hybrid computing. And it would not be just infrastructure driving this growth, but also the intelligence we would infuse into applications. We would offer cognitive services for vision, speech, text, recommendations, and face and emotion detection. Developers would simply use APIs within their applications to augment users' experiences by enabling solutions to see, hear, speak, and interpret the world around them. Our intelligent cloud would democratize these capabilities for startups, small businesses, and enterprises alike.

Third, we needed to move people from needing Windows to choosing Windows to loving Windows by creating more personal computing. Just as we would transform business and society through cloud computing, we also needed to revolutionize the workplace to help organizations and people be more productive. We launched Windows 10 with a new concept—to enable Windows as a service, continuously delivering value across all of our products. We engineered Windows 10 to enable innovative and more natural ways to interact and engage with devices—ask a question with your voice, draw with the flick of a pen, and secure your most important things with a smile or a touch. These experiences place users at the center so they can move seamlessly across all devices—from the PC, Xbox, phones, and Surface Hub, to Microsoft HoloLens, and Windows Mixed Reality.

We needed employees and partners on board for the transformation ahead, and we needed Wall Street to be with us as well. Amy Hood, our CFO, understood the culture change we needed to navigate. She also became the crucial partner I needed for precise attention to quantitative detail across the business. Her job is where the rubber meets the road. Ahead of my first financial analyst meeting, Amy helped to translate the mission and ambitions into language and goals investors needed to hear. She helped, for example, shape the goal to build a $20 billion cloud business, something investors grabbed on to and tracked quarter after quarter. It took us from a defensive frame amid falling PC and phone share to an offensive mindset. We went from deflection to ownership of our future.

Rediscovering the soul of Microsoft, redefining our mission, and outlining the business ambitions that would help investors and customers grow our company—these had been my priorities with the first inkling that I would become CEO. Getting our strategy right had preoccupied me from the beginning. But as management guru Peter Drucker once said, "Culture eats strategy for breakfast." As I concluded my talk that morning in Orlando, I focused on what would be our grandest endeavor, the highest hurdle—transforming the Microsoft culture.

—

It's surprising when an arena jammed to the rafters with fifteen thousand people falls silent. It's also unsettling when nothing can be seen because of the blinding stage lights. That's how I felt as I stood onstage in Orlando. I could feel a small lump grow in my throat. I was about to launch into a topic that was at once crucial for Microsoft to get right, but also deeply personal for me.

"I'm going to close out by talking about our culture. To me it is everything," I said.

Bill and Steve had made this annual address many times to employees over the years. Bill often looked off into the future, predicting tech trends and how Microsoft would lead. Steve rallied the troops, whipping everyone into a frenzy of excitement. I had used the first part of my speech to proclaim a new mission, one rooted in rediscovering the soul of our company. I had outlined a series of new business ambitions. But as I had foreshadowed in that Thanksgiving memo to the board of directors, real change depended on culture change.

Culture can be a vague and amorphous term. In his perceptive book, *Culture*, the literary theorist Terry Eagleton wrote that the idea of culture is multifaceted, "a kind of social unconscious." With razor precision, he separates culture into four different meanings, but the most relevant for an organization is the values, customs, beliefs, and symbolic practices that men and women live and breathe each day. Culture is made up of acts that become habitual and accrue to something coherent and meaningful. Eagleton, who lives in Ireland, notes that a mailbox in his country is evidence of civilization, but the fact they are all painted green is evidence of culture. I think of culture as a complex system made up of individual mindsets—the mindsets of those in front of me. Culture is how an organization thinks and acts, but individuals shape it.

In my own life, it's the language, routines, and mindset of my parents back in India and my immediate family in Seattle that helped form me and still guide me to this day. It's that diverse collection of classmates back in Hyderabad who shared a learning mindset that would propel them on to leadership in government, business, sports, and entertainment. In all of these experiences, I'd been encouraged to follow my curiosity and to push the limits of my own capabilities, and now I was beginning to see how this approach would be critical to Microsoft as it confronted the burden of its past success.

Earlier in the year, Anu had handed me a copy of Dr. Carol Dweck's book, *Mindset: The New Psychology of Success*. Dr. Dweck's research is about overcoming failures by believing you can. "The view you adopt for yourself profoundly affects the way you lead your life." She divides the world between learners and non-learners, demonstrating that a fixed mindset will limit you and a growth mindset can move you forward. The hand you are dealt is just the starting point. Passion, toil, and training can help you to soar. (She even writes persuasively about what she calls the "CEO disease," an affliction of business leaders who fail to have a growth mindset.)

My wife wasn't thinking of my success when she gave me Dr. Dweck's book. She was thinking of the success of one of our daughters who has learning differences. Her diagnosis took us on a journey of discovery to help her. First was the internal journey, concern for her but also the need to educate ourselves. Next came action. We found a school in Vancouver, Canada, that specializes in learning differences like hers. We spent five years of our lives splitting time and family between Vancouver and Seattle in order to augment her regular schooling while keeping Zain's care consistent in Seattle.

All of this meant separation at many levels: husband and wife; father and daughters; mother and son. We were maintaining two lives in two countries. Anu drove thousands of miles between Seattle and Vancouver in rain, snow, and darkness, and so did I on alternate weekends for five years. It was a trying time, but Anu and the girls made some exceptional friends in Canada. As a family, we learned together that these predicaments were universal. Families from California, Australia, Palestine, and New Zealand converged on the Vancouver school with issues and challenges. I discovered that recognition of these universal predicaments leads to universal empathy—empathy for and among children, adults, parents, and teachers. Empathy, we learned, was indivisible and was a universal value. And we learned that empathy is essential to deal with problems everywhere, whether at Microsoft or at home; here in the United States or globally. That is also a mindset, a culture.

As I continued my speech at the global sales conference, the empathy I felt for my kids and the empathy I felt for the people listening in that audience were on my mind and in my emotions.

"We can have all the bold ambitions. We can have all the bold goals. We can aspire to our new mission. But it's only going to happen if we live our culture, if we teach our culture. And to me that model of culture is not a static thing. It is about a dynamic learning culture. In fact, the phrase we use to describe our emerging culture is 'growth mindset,' because it's about every individual, every one of us having that attitude—that mindset—of being able to overcome any constraint, stand up to any challenge, making it possible for us to grow and, thereby, for the company to grow."

I told my colleagues that I was not talking bottom-line growth. I was talking about our individual growth. We will grow as a company if everyone, individually, grows in their roles and in their lives. My wife, Anu, and I had been blessed with wonderful children, and we've had to learn their special needs. That has changed everything for us. "It's taken me on this journey of developing more empathy for others. And what gives me deep meaning is that ability to take new ideas and empathy for people; to connect the two and have great impact. That's what gives me the greatest satisfaction. It's why I work for Microsoft. And that's what I aspire for each one of you to do as you work here."

Our culture needed to be about realizing our personal passions and using Microsoft as a platform to pursue that passion. For me, my greatest satisfaction has come from my passion to see technology become more accessible for people with disabilities and to help improve their lives in a myriad of ways.

Just as my predecessor Steve Ballmer had done at these annual gatherings, I'd closed my speech with a call to action, but one with a very different mood and purpose. I had essentially asked employees to identify their innermost passions and to connect them in some way to our new mission and culture. In so doing we would transform our company and change the world. When you're CEO, these goals can be easy to imagine, but when an employee's aperture is smaller—a marketer in Malaysia or technical support in Texas—such a mission can seem distant and unattainable. So the challenge I'd set forth in my speech might be a daunting one. I wondered whether I'd connected with the audience or left them baffled and untouched.

Feeling my emotions beginning to overcome me, I skipped my last slide and quickly exited the stage. Jill pointed at the doorway to the auditorium, not my private green room, "Watch with them." As a video started presenting not just the year's progress but the expansive, mission-driven opportunity ahead, I slipped back into the auditorium through a side entrance. No one could see me in the darkened auditorium. Every eye was glued to the screen, but I was watching them, gauging the emotion in the room. Everyone was locked in and some were softly wiping away tears. I knew then that we were onto something.

CHAPTER 4

A Cultural Renaissance

From Know-It-Alls to Learn-It-Alls

A few days later, I was in Nanyuki, Kenya, standing inside a solar-powered shipping container that doubles as an Internet café. One of our partners, Mawingu (Swahili for "cloud") Networks, provides these rural communities with low-cost Internet services, which has opened up access to knowledge for schoolkids and parents alike. In fact, in just one short year educational test scores have dramatically increased.

Inside the café, I stopped to chat with Chris Baraka, who uses the connectivity here to make a living as a writer and teacher. I also observed farmers pausing from their outdoor work to check crop prices. With only a dozen people in attendance, one would never know that I was celebrating the global launch of Windows 10, a Microsoft product at the core of our strategy.

Two decades earlier, the launch of Windows 95, with its pricey theme song, "Start Me Up" by the Rolling Stones, midnight retail parties, and media hysteria, had helped create a phenomenon of bigger and bigger software launches. Competitors tried to outdo and out-lavish each other's launch events in hopes of sparking customer impulses to purchase. But that was then, and this is now. A refreshed product launch strategy, one that reflected the times and our new mission and culture, was needed. Initially our communications chief, Frank Shaw, had presented a first-class launch plan complete with visual extravaganzas like setting ablaze the Sydney Opera House with Windows-branded colorful lights. He felt the product needed these kinds of inspiring and hopefully media-attention-getting images in Paris, New York, Tokyo, and elsewhere in order to generate the news coverage we had come to expect. But the approach didn't feel right. I felt that this was one of those moments to show a different Microsoft. We were struggling with what to do, and I decided to take a brief break from the meeting to get coffee. During that break, a side conversation among some of the team members rose above the chatter. "We should launch Windows 10 in Kenya." Kenya is a country where we have customers, partners, and employees. It's a nation with great promise, one that with digital transformation could leapfrog over others by getting infrastructure and skills in place.

The launch of Windows 10 wasn't about one product; it was about our mission. And if we're going to seek to empower every person on the planet, why not go to the other side of the planet and make that real? I walked down the hall to Frank's office. "Let's take a chance." I knew that we had built a low-cost, high-speed Internet connectivity solution that leverages an innovative technology called "TV white space"—the unused broadband spectrum that exists in between television channels—to connect rural, poor areas like Nanyuki, Kenya, to the Web. We could show off not just Windows 10 but also its relevance to anyone and everyone, no matter their geography or socioeconomic status. Frank thought about it another minute and agreed. What better way to demonstrate our new mission and emerging culture than to be in eastern Africa, a region that exemplifies both the challenges and the opportunities for technology to transform and create economic growth? We might not get all of the TV cameras we used to get, but we'd be demonstrating our desire to understand every customer's context, including that of farmers in a remote African village, for whom technology tools can mean the difference between abject poverty and hope. By embracing this new cultural mindset, we'd be able to start listening, learning more and talking less.

So what did this growth mindset reveal? One of the lessons we took back is that it's too simplistic to call a country like Kenya a developing economy or the United States a developed one. Both countries have educated, tech-savvy customers capable of using our most sophisticated products, and both countries have potential customers with little or no skills. Sure, there are higher concentrations of one or the other in each country, but it's a false distinction simply to think of countries as either developed or developing. The Windows 10 launch in Kenya struck a far more global tone for the company, and it also taught us valuable lessons.

I like to think that the *C* in CEO stands for culture. The CEO is the curator of an organization's culture. As I had told employees in Orlando, anything is possible for a company when its culture is about listening, learning, and harnessing individual passions and talents to the company's mission. Creating that kind of culture is my chief job as CEO. And so, whether it was through public events like the launch of Windows 10 or through speeches, emails, tweets, internal posts, or monthly employee Q&A sessions, I planned to use every opportunity at my disposal to encourage our team to live this culture of dynamic learning.

Of course, exhortations from the CEO are only a fraction of what it takes to create real culture change, especially in a huge, very successful organization like Microsoft. An organizational culture is not something that can simply unfreeze, change, and then refreeze in an ideal way. It takes deliberate work, and it takes some specific ideas about what the culture should become. It also requires dramatic, concrete actions that seize the attention of team members and push them out of their familiar comfort zones.

Our culture had been rigid. Each employee had to prove to everyone that he or she knew it all and was the smartest person in the room. Accountability—delivering on time and hitting numbers—trumped everything. Meetings were formal. Everything had to be planned in perfect detail before the meeting. And it was hard to do a skip-level meeting. If a senior leader wanted to tap the energy and creativity of someone lower down in the organization, she or he needed to invite that person's boss, and so on. Hierarchy and pecking order had taken control, and spontaneity and creativity had suffered as a result.

The culture change I wanted was actually rooted in the Microsoft I originally joined. It was centered on exercising a growth mindset every day in three distinct ways.

First, we needed to obsess about our customers. At the core of our business must be the curiosity and desire to meet a customer's unarticulated and unmet needs with great technology. There is no way to do that unless we absorb with deeper insight and empathy what they need. To me this was not something abstract, but rather something we all get to practice each day. When we talk to customers, we need to listen. It's not an idle exercise. It is about being able to predict things that customers will love. That's growth mindset. We learn about our customers and their businesses with a beginner's mind and then bring them solutions that meet their needs. We need to be insatiable in our desire to learn from the outside and bring that learning into Microsoft, while still innovating to surprise and delight our users.

Second, we are at our best when we actively seek diversity and inclusion. If we are going to serve the planet as our mission states, we need to reflect the planet. The diversity of our workforce must continue to improve, and we need to include a wide range of opinions and perspectives in our thinking and decision making. In every meeting, don't just listen—make it possible for others to speak so that everyone's ideas come through. Inclusiveness will help us become open to learning about our own biases and changing our behaviors so we can tap into the collective power of everyone in the company. We need not just value differences but also actively seek them out, invite them in. And as a result, our ideas will be better, our products will be better, and our customers will be better served.

Finally, we are one company, one Microsoft—not a confederation of fiefdoms. Innovation and competition don't respect our silos, our org boundaries, so we have to learn to transcend those barriers. We are a family of individuals united by a single, shared mission. It is not about doing what's comfortable within our own organization, it's about getting outside that comfort zone, reaching out to do things that are most important for customers. For some companies this comes more naturally. For example, those tech companies born with an open-source mentality get it. One group may create code and intellectual property but it's open and available for inspection and improvement from other groups inside and outside the company. I tell my colleagues they get to own a customer scenario, not the code. Our code may need to be tailored one way for a small business and another way for a public-sector customer. It's our ability to work together that makes our dreams believable and, ultimately, achievable. We must learn to build on the ideas of others and collaborate across boundaries to bring the best of Microsoft to our customers as one—one Microsoft.

When we exercise a growth mindset by being customer-obsessed, diverse, and inclusive and act as One Microsoft, that's when we live our mission and truly make a difference in the world. Taken together, these concepts embody the growth in culture I set out to inculcate at Microsoft. I talked about these ideas every chance I got. And I looked for opportunities to change our practices and behaviors to make the growth mindset vivid and real. Part of the culture change was to give people the breathing room, the space, to bring their own voices and experiences to the conversation. The last thing I wanted was for employees to think of culture as "Satya's thing." I wanted them to see it as their thing, as Microsoft's thing.

To encourage the shift toward a learning culture, we created an annual hackathon during our OneWeek celebration, a time for everyone to be on campus simultaneously to make connections, learn about what others are doing, find inspiration, and collaborate. Playing off the notion of growth mindset, the hack made perfect sense. Within the subculture of computer programmers, hacking is a time-honored tradition of working around limitations and creatively solving a difficult problem or opportunity. In that first year, more than twelve thousand employees from eighty-three countries entered more than three thousand hacks ranging from ending sexism in video games to making computing more accessible to people with disabilities to improving industrial supply-chain operations.

One team was made up of people from multiple Microsoft groups across the company. They were interested in delivering better learning outcomes for kids with dyslexia. The Microsoft hackathon became an avenue for people with depth and passion, people spanning product groups like OneNote and Windows as well as research to come together and start a movement. They began by researching the science surrounding dyslexia and decided to go after a problem called visual crowding. Led by one of our software engineers, the team found ways to allow more space in between letters to make words more readable. But they didn't stop there. They also found ways to create a more immersive reading function with the ability to highlight text and have it read out loud, further increasing reading comprehension. They built a tool to break words into syllables and to highlight the verb and subordinate clause. They got feedback from students and teachers. In fact, one teacher wrote to tell us about the gains she had seen in her classroom, including from a boy with dyslexia who could only read six words per minute. Even when he did make a fluency gain, he couldn't sustain it. When he started using the tools our team built, she saw an immediate change. He was more willing to attempt assignments and his reading fluency skyrocketed. He went from reading only six words per minute to twenty-seven words per minute in a matter of weeks. Another student improved so much he was moved to a higher-level reading class. Today, the functionality that began as a Hackathon project is now built into some of our most important products, including Word, Outlook, and the Edge browser.

Now the annual growth hack has become a Microsoft tradition. Every year, employees—engineers, marketers, all professions—prepare in their home countries for the OneWeek growth hack like students preparing for a science fair, working in teams to hack problems they feel passionate about and then developing presentations designed to win votes from their colleagues. Gathered in tents named Hacknado and Codapalooza, they consume thousands of pounds of doughnuts, chicken, baby carrots, energy bars, coffee, and the occasional beer to fuel their creativity. Programmers and analysts suddenly transform into carnival barkers, selling their ideas to anyone who will listen. Reactions range from polite questions to vigorous debate and challenges. In the end, votes sent from smartphones are tallied, projects evaluated, winners celebrated. A few projects even receive funding as new business efforts.

Because I've made culture change at Microsoft such a high priority, people often ask how it's going. Well, I suppose my response is very Eastern: We're making great progress, but we should never be done. It's not a program with a start and end date. It's a way of being. Frankly, I am wired that way. When I learn about a shortcoming, it's a thrilling moment. The person who points it out has given me the gift of insight. It's about questioning ourselves each day: Where are all the places today that I had a fixed mindset? Where did I have a growth mindset?

As CEO, I'm not exempt from having to ask myself these questions. Each of my business decisions can be scrutinized in terms of whether or not it has helped Microsoft shift toward the growth mindset we aspire to.

Fixed-mindset decisions are ones that reinforce the tendency to continue doing what we've always done. Traditionally, when we launched a new version of Windows, existing Windows users would pay us to upgrade. Terry Myerson, the executive in charge of our Windows and devices group, had the growth mindset to shift, for a time, to a free consumer upgrade and forgo that revenue. In just a little over a year it had become the most popular Windows upgrade ever with hundreds of millions of users and still rising. We wanted customers to make that shift to loving Windows and to have the most personal and secure devices.

Upon reflection, we learned a lot from Nokia, even though it resulted in a painful write-down of the assets. Acquiring the Finnish smartphone company led to numeric growth in terms of people and revenue, but ultimately we failed to break through in the highly competitive mobile phone business. Importantly, though, we learned a lot about what it means to design, build, and manufacture hardware.

Our acquisition of Sweden-based Mojang and its video game Minecraft also represented a growth mindset because it created new energy and engagement for people on our mobile and cloud technologies, and it would open new opportunities in the education software space.

The story of how the Minecraft acquisition happened illustrates some of the key qualities of a growth mindset, including the readiness to empower and learn from individuals who possess insights and passion that the rest of the organization needs to learn from. In this case, the individual was Phil Spencer, who heads Xbox. Phil understood that we needed to be the most attractive platform in the world for gamers, and he knew Minecraft had a dedicated and gigantic community of players who invented and built new worlds in this virtual Lego-like video game.

It's the rare video game that is invited into the classroom, and Minecraft is not just invited but desired. Teachers love the way it encourages building, collaboration, and exploration. It's a 3D sandbox of sorts. If the classroom curriculum calls for building a river ecosystem with marshes, Minecraft can do that. If the river needs to flow, the Minecraft logic function can make that happen. It teaches digital citizenship because it's multiplayer. Twelve students in a classroom can be told to go build a house and within minutes they form teams and get to work—a model of the workplace of the future.

Phil and his team built a great relationship with the Swedish game studio and managed to expand the Minecraft franchise to multiple devices including mobile and console. Early in Microsoft's relationship with Mojang, before I was CEO, Phil presented an opportunity to purchase Minecraft, but Phil's boss at the time chose not to move forward. For some, such a visible, high-level rejection could have been withering, but Phil didn't give up. He knew that this beloved game belonged in a place where it could continue to scale up and prosper. He also knew that for Microsoft, bringing Minecraft into our ecosystem could lead to deeper engagement with the next generation of gamers. He knew our cloud could help it scale to reach every corner of the globe.

Phil maintained a great relationship with Mojang, continuing to build trust, and one day, Phil's team got a text that the company was for sale again. They could have gone to any of our competitors to strike a deal, but they came back to us. Phil had recently become the head of Xbox, and I was new in my role as CEO. He brought the deal to me for reconsideration. I felt we could bring the inherent strengths of Microsoft to the product while preserving the integrity and creativity of the small indie group that invented it. We pulled the trigger on a $2.5 billion acquisition. Today Minecraft is one of the bestselling games of all time on the PC, Xbox, and mobile. It has tremendous and lasting gamer engagement. Bill Gates and Steve Ballmer, who were still on the board when the deal was presented, later laughed and said they had initially scratched their heads, failing to understand the wisdom of the move. Now we all get it.

That's a growth mindset, and it highlights individual empowerment—what one person or one team can do against the odds.

Even though I am ambivalent about questions from outsiders about how the culture change is going, it's easy to see there is a tangible shift happening inside Microsoft. If you want to understand the culture inside a software company, show up at a meeting that includes engineers from different parts of the company. These are very smart people who are passionate about building great products. But are they plugged into what customers need and want? Do they include diverse opinions and capabilities when writing code? And do they act like they're on the same team, even if they work in different groups? Answers to questions like these serve as a great barometer for the culture we need. Demonstrating a growth mindset. Customer-centric. Diverse and inclusive. One company.

I remember a gathering in 2012 of top engineers from across Microsoft. It was one of a series of WHiPS, short for Windows High-Powered Summits, envisioned as opportunities to improve products and solve problems that rely on collaboration across code bases. There was a high degree of ownership and a lot of pride. But to my dismay, the meeting deteriorated into a gripe session. One developer argued that he had fixed something in the Windows code base that would help with a problem customers had discovered in an application that ran on Windows. Even though he had fixed it, the Windows developers were not accepting, or "checking-in," his new code. The discussion quickly devolved into argument and then name-calling. This was not the culture we're looking for.

When I attended another WHiPS a few years later, I heard a very different conversation. A developer announced that he had found a means for capturing, or taking a screenshot of, a moving image—a big improvement over our existing "snipping" tool, which was capable only of capturing a static image. A small piece of code that makes a big difference to a designer or editor. As in 2012, though, his fix had not *yet* been integrated into the Windows code. With a growth mindset, *yet* is an important clarification.

Terry Myerson, the head of Windows, jumped into the conversation before the arguing and finger-pointing could begin. "Send the fix again and we'll take care of it."

Even back in 2012 the energy for culture change had been there, but we needed to create a conduit for change. We had to break down the dam and allow change to flow. Now that has begun to happen.

The key to the culture change was individual empowerment. We sometimes underestimate what we each can do to make things happen, and overestimate what others need to do for us. We had to get out of the mode of thinking in which we assume that others have more power over us than we do. I became irritated once during an employee Q&A when someone asked me, "Why can't I print a document from my mobile phone?" I politely told him, "Make it happen. You have full authority."

Another time, members of a chat group on Yammer, our corporate social media service for internal conversations, were complaining that people were leaving half-used milk cartons in the office refrigerator. Apparently people would open a fresh eight-ounce container of milk, pour some in their coffee or tea, and then leave it out on the counter thinking others would finish it. But no one wants to use a personal milk container opened by somebody else that is beginning to sour. It blew up on Yammer, and I used one of my video messages to employees to have a good laugh at it, showcasing it as a humorous example of a fixed mindset.

Culture change is hard. It can be painful. The fundamental source of resistance to change is fear of the unknown. Really big questions for which there are no certain answers can be scary.

Consider one of the questions we ask ourselves continually: What is the computing platform of the future? Windows has been the PC platform of choice for decades, but now we're imagining a new era. The cloud and its edge with multisensory and multi-device experiences will enable new computers and new computing that is sensitive to human presence and responsive to individual preferences. We're hard at work building the ultimate computing experience, blending mixed reality, artificial intelligence, and quantum computing. Which of these will dominate the computing world of 2050—or will some new breakthrough emerge that is currently unimagined?

Anyone who says they can accurately predict the future trajectory of tech is not to be trusted. However, a growth mindset enables you to better anticipate and react to uncertainties. Fear of the unknown can send you in a million directions, and sometimes it just dead-ends with inertia. A leader has to have an idea what to do—to innovate in the face of fear and inertia. We need to be willing to lean into uncertainty, to take risks, and to move quickly when we make mistakes, recognizing failure happens along the way to mastery. Sometimes it feels like a bird learning to fly. You flap around for a while, and then you run around. Learning to fly is not pretty but flying is.

If you want to see what flapping around looks like, do a search for me and karma. It's a fall day in Phoenix, Arizona, and I am attending the Grace Hopper celebration of women in computing, the world's largest gathering of women technologists. Diversity and inclusion is a bedrock strategy in building the culture we need and want, but as a company and as an industry we've come up far too short. According to one report, women in the United States held 57 percent of professional occupations in the 2015 workforce, but only 25 percent of professional computing occupations. That's a real problem, and one that will only get worse with inaction since the number of computer-related jobs is only increasing. As the leader of a company, a husband and the father of two young women, I see this failure to attract and retain women in computing as bad business, and it's wrong. Which makes what I said that day in Phoenix all the more perplexing, not to mention embarrassing. Early in my appearance there were loud cheers when I said we cannot settle for supply-side excuses. The real issue is how to get more women inside the organization. But near the end of my interview onstage, Dr. Maria Klawe, a computer scientist, president of Harvey Mudd College and a former Microsoft board member, asked me what advice I had for women seeking a pay raise who are not comfortable asking. It's a great question because we know women leave the industry when they are not properly recognized and rewarded. I only wish my answer had been great. It was not. I paused for a moment and remembered an early president at Microsoft who had told me once that human resource systems are long-term efficient but short-term inefficient. In other words, over time you are rewarded and recognized for stellar work but not always in real time. "It's not really about asking for the raise but knowing and having faith that the system will actually give you the right raises as you go along," I responded. "And that might be one of the additional superpowers that women who don't ask for the raise have, because that's good karma. It'll come back. Long-term efficiency solves it." Dr. Klawe, whom I respect enormously, kindly pushed back. "This is one of the few things I disagree with you on," eliciting scattered applause from the audience. She used it as a teaching moment, directing her comments to the women in the audience but clearly giving me a lesson I won't forget. She told the story of a time when she was asked how much pay would be sufficient, and she just said whatever is fair. By not advocating for herself, she didn't get what was fair. Having learned that lesson the hard way, she encouraged the audience to do their homework and to know what the proper salary is. Afterward, we hugged and left the stage to warm applause. But the damage was done. The criticism, deserved and biting, came swiftly through waves of social media and international radio, TV, and newspaper coverage. My chief of staff smugly read me a tweet capturing the moment, "I hope Satya's comms person is a woman and is asking for a raise right now."

Honestly, I left the conference inspired and energized, but I was mad at myself for blundering such an important chance to communicate my own commitment and Microsoft's to increasing the number of women we hire at every level of our industry. I was frustrated, but I also was determined to use the incident to demonstrate what a growth mindset looks like under pressure. A few hours later I shot off an email to everyone in the company. I encouraged them to watch the video, and I was quick to point out that I had answered the question completely wrong. "Without a doubt I wholeheartedly support programs at Microsoft and in the industry that bring more women into technology and close the pay gap. I believe men and women should get equal pay for equal work. And when it comes to career advice on getting a raise when you think it's deserved, Maria's advice was the right advice. If you think you deserve a raise, you should just ask." A few days later, in my regular all-employee Q&A, I apologized, and explained that I had received this advice from my mentors and had followed it. But this advice underestimated exclusion and bias—conscious and unconscious. Any advice that advocates passivity in the face of bias is wrong. Leaders need to act and shape the culture to root out biases and create an environment where everyone can effectively advocate for themselves. I had gone to Phoenix to learn, and I certainly did. But perhaps what taught me more was hearing stories from women I deeply respect about the bias they experienced earlier in their careers: being told to smile more often, being blocked from joining the good old boys' club, or facing the difficult trade-off between taking time off after having a baby or relentlessly climbing the career ladder—these powerful women who shared the hurt of their past experiences with me. During this time, I also found myself reflecting on the sacrifices my mother made for me and the challenging decision Anu had made to leave her promising career as an architect to care for Zain and our two girls full-time for more than two decades. She made it possible for my career to advance at Microsoft.

Since my remarks at Grace Hopper, Microsoft has made the commitment to drive real change in this area—linking executive compensation to diversity progress, investing in diversity programs, and sharing data publicly about pay equity for gender, racial, and ethnic minorities. In some ways, I'm glad I messed up in such a public forum because it helped me confront an unconscious bias I didn't know I had, and it helped me find a new sense of empathy for the great women in my life and at my company.

This episode led me to reflect on my own experience as an immigrant. Hearing racial slurs toward Indians after moving to America never stung me, I just blew them off—an easy thing to do for a man raised in the majority and with privilege in India. Even when some people in positions of power have remarked that there are too many Asian CEOs in technology, I've ignored their ignorance. But as I grow older, and watch a second generation of Indians—my kids and their friends—grow up as minorities in the United States, I cannot help but think about how our experiences differ. It infuriates me to think they will hear and grapple with racial slurs and ignorance.

When I joined Microsoft, there was an undercurrent among the Indian engineers and programmers. We were aware that despite our contributions, there had yet to be one of us promoted to vice president, a rank that recognizes a leader as an officer of the company. We could get to a certain level but not beyond. In fact, a senior executive, long since gone from the company, once told another Indian colleague that it was because of our accents—an idea as derogatory as it is outdated. It was the 1990s and I was surprised to hear such bias within such a leading-edge company, especially one led by and founded by such open-minded leaders. Yet, when I looked around, sure enough, there were no Indian VPs despite the well-known top performance of so many Indian engineers and managers. It was not until 2000 that myself and a few other Indians were promoted to the executive ranks.

Whether cultural or attained wisdom, we felt that if we worked hard and kept our heads down, eventually good things would happen. One of my colleagues at the time, Sanjay Parthasarathy, became a big influence in my life and career. Although we had not known each other in India, Sanjay did play cricket under my own school captain for South Zone in U-19s. At Microsoft, we hit it off immediately. The combination of cricket and technology meant we never ran out of things to talk about. He told me that

I must internalize for myself the belief that the sky is the limit. I must work hard—not to climb the ladder, but to do important work. With benefit of hindsight, I know now that anyone who feels like an outsider can be successful, but it requires both an enlightened management and a dedicated employee. A manager can be demanding, but must also have the empathy to figure out what will motivate employees. Likewise, an employee is right to put his or her head down and work hard, but they also have the right to expect a pathway to greater responsibility and recognition when they do. There must be balance.

Because of my own experience, and by learning from my colleagues, I get now how hard it is to join a company that doesn't look like you and to live in a community where most of your neighbors don't look like you. How do you identify role models you can fully relate to? How do you find mentors, coaches, and sponsors who can help you succeed without hiding your true self? At work, the tech industry, including Microsoft, is simply not as diverse as we must become. And outside work, minorities can also feel isolated. King County in Washington state, for example, which encompasses Redmond, Bellevue, and Seattle, is 70 percent white. African Americans comprise under 7 percent and Latinos and Hispanics are nearly 10 percent. To help connect communities of people with like backgrounds and interests, there has been a long tradition inside the company of underrepresented groups organizing themselves into employee resource groups such as Blacks @ Microsoft (BAM) and Women @ Microsoft. In all there are seven major ERGs and forty more specific networks. They host online discussions, networking meetings, provide mentoring and professional development, community outreach, and connect people with a community inside and outside of work. Most importantly, they offer support. During 2016, as our African American colleagues struggled to come to terms with the tragic episodes of violence and innocent lives lost here in the United States, the BAM community was a source of connection and support. Following the Orlando nightclub massacre, the email discussion group for GLEAM, the employee resource group for the LGBTQ community at Microsoft, provided a much-needed safe space for members to air fears and concerns. We all want a culture in which we're heard and supported.

—

I said earlier that culture can be a vague, amorphous term. That's why we worked so carefully to define the culture we wanted. And it's why we measure everything. When it comes to humans, data is not perfect, but we can't monitor what we can't measure. So, we routinely survey employees to take their pulse.

After three years of intensive focus on culture-building we began to see some encouraging results. Employees told us they felt the company was heading in the right direction. They felt we were making the right choices for long-term success, and they saw different groups across the company working together more. This was exactly what we were hoping for.

But we also saw some trends that were not as encouraging. When asked whether their vice president, or group leader, was prioritizing talent movement and development, the results were worse than they'd been before our culture-building project began. Even the most optimistic workers will become discouraged if they are not being developed. I had set a clear mission and envisioned an empowering culture. Employees and senior leaders were on board, but we had a missing link—middle management.

This was a bit disheartening, but in retrospect it was completely understandable. Remember what I said about the aperture of those colleagues sitting in the darkened theater in Orlando. The opening through which a middle manager can see the organizational culture in the midst of his or her daily work is a crack when compared with the panoramic view a CEO enjoys. A *Harvard Business Review* survey found that senior leaders inside companies spend less than 10 percent of their time developing high potential leaders. If even top executives cannot find the time to unlock employee potential, the growth path for most corporate team members looks pretty static.

After reviewing the results, I seized on an upcoming meeting with about 150 of my top leaders to tell a few stories and to share my expectations with them. First off, I told them about an anonymous Microsoft manager who had come to me recently to share how much he loved the new growth mindset and how much he wanted to see more of it pointing out, "Hey, Satya, I know these five people who don't have a growth mindset." The guy was just using growth mindset to find a new way to complain about others. That is not what we had in mind.

I told these high-potential leaders that once you become a vice president, a partner in this endeavor, the whining is over. You can't say the coffee around here is bad, or there aren't enough good people, or I didn't get the bonus.

"To be a leader in this company, your job is to find the rose petals in a field of shit."

Perhaps not my best line of poetry, but I wanted these people to stop seeing all the things that are hard and start seeing things that are great and helping others see them too. Constraints are real and will always be with us, but leaders are the champions of overcoming constraints. They make things happen. Every organization will say it differently, but for me there are three expectations—three leadership principles—for anyone leading others at Microsoft.

The first is to bring clarity to those you work with. This is one of the foundational things leaders do every day, every minute. In order to bring clarity, you've got to synthesize the complex. Leaders take internal and external noise and synthesize a message from it, recognizing the true signal within a lot of noise. I don't want to hear that someone is the smartest person in the room. I want to hear them take their intelligence and use it to develop deep shared understanding within teams and define a course of action.

Second, leaders generate energy, not only on their own teams but across the company. It's insufficient to focus exclusively on your own unit. Leaders need to inspire optimism, creativity, shared commitment, and growth through times good and bad. They create an environment where everyone can do his or her best work. And they build organizations and teams that are stronger tomorrow than today.

Third, and finally, they find a way to deliver success, to make things happen. This means driving innovations that people love and are inspired to work on; finding balance between long-term success and short-term wins; and being boundary-less and globally minded in seeking solutions.

I love these three leadership principles. The heart of my message: Changing the culture at Microsoft doesn't depend on me, or even on the handful of top leaders I work most closely with. It depends on everyone in the company—including our vast cadre of middle managers who must dedicate themselves to making everyone they work with better, every day.

I totally empathize with other leaders, and see my job as helping them become even better. Leadership can be a lonely business. It can also be a noisy place. When a leader steps into the arena, especially in today's loud echo chamber of social media, he or she can be tempted to make decisions that will result in instant gratification. But we have to look beyond the temporal, discounting what someone will write in this moment's tweet or tomorrow's news. Reasoned judgment and inner conviction are what I expect from myself and from the leaders around me. Make the call, but don't expect consensus.

Internally, we needed to have strong partnerships—between leaders across the company among teams. But that same growth mindset was needed externally, too. The competitive landscape had shifted seismically over the previous decade, and now new and surprising partnerships with friends and former enemies were needed.

CHAPTER 5

Friends or Frenemies?

Build Partnerships Before You Need Them

There was an audible gasp and more than a smattering of chuckles in the auditorium when I reached into my suit jacket and pulled out an iPhone. No one in recent memory had seen a Microsoft CEO showing off an Apple product. Especially not at a competitor's sales conference.

"This is a pretty unique iPhone," I told attendees at Salesforce's annual marketing event as the crowd quieted down. Salesforce both competes and partners with Microsoft in online services. "I like to call it the iPhone Pro because it's got all the Microsoft software and applications on it."

On the giant screen behind me, a close-up of the phone appeared. One by one, the app icons flashed into view—iPhone versions of Microsoft classics like Outlook, Skype, Word, Excel, and PowerPoint as well as newer mobile applications like Dynamics, OneNote, OneDrive, Sway, and Power BI. The crowd erupted in applause.

Seeing me demo Microsoft software on an iPhone designed and built by Apple, one of our toughest, longest-standing competitors, was surprising and even refreshing. Microsoft versus Apple has been such a prominent and even contentious rivalry that people forget we've been building software for the Mac since 1982. Today one of my top priorities is to make sure that our billion customers, no matter which phone or platform they choose to use, have their needs met so that we continue to grow. To do that, sometimes we have to bury the hatchet with old rivals, pursue surprising new partnerships, and revive longstanding relationships. Over the years we've developed the maturity to become more obsessed with customer needs, thereby learning to coexist *and* compete.

In the previous chapter I wrote about the importance of building the right culture. Healthy partnerships—often difficult but always mutually beneficial—are the natural and much-needed product of the culture we're building. Steve Ballmer helped me deeply understand this with his three Cs. Imagine a target with three concentric rings. The outer ring is *concepts*. Microsoft, Apple, or Amazon may have an exciting product idea, but is that enough? An organization may have a conceptual vision—a dream or imagination filled with new ideas and new approaches, but do they have what's in the second ring: *capabilities*? Do they have the engineering and design skills required to actually build that concept alone? And finally, the bull's-eye, is a *culture* that embraces new concepts and new capabilities and doesn't choke them out. That's what's needed in order to build and sustain innovation-producing and customer-pleasing products—smart partnerships. Concepts are better and capabilities more comprehensive when the culture invites partners to the table. Two or more heads really are better than one.

A few years back, Apple had a concept they felt would benefit from a renewed partnership with our capabilities and culture. Shortly after becoming CEO I decided we needed to get Office everywhere, including iOS and Android. We had these versions in the works for some time, just waiting for the right moment to launch. I wanted unambiguously to declare, both internally and externally, that the strategy would be to center our innovation agenda around users' needs and not simply their device. We announced that we would bring Office to iOS in March 2014, two months into my new role.

Soon Apple sent a cryptic note to our Office team asking for an engineer to sign a nondisclosure agreement and come to Cupertino for a meeting. This is standard operating procedure in our secretive industry where intellectual property must be guarded. After a few meetings, it became clear that what Apple wanted was for Microsoft to work with them to optimize Office 365 for their new iPad Pro. Apple told us that they felt there was a new openness at Microsoft. They trusted us and wanted us to be part of their launch event.

There was passionate debate internally about whether this was a good idea. At first some product-line leaders within Microsoft felt uneasy about partnering with their competitor; I definitely heard some resistance behind closed doors. One way to explain the logic is by turning to game theory, which uses mathematical models to explain cooperation and conflict. Partnering is too often seen as a zero-sum game—whatever is gained by one participant is lost by another. I don't see it that way. When done right, partnering grows the pie for everyone—for customers, yes, but also for each of the partners. Ultimately the consensus was that this partnership with Apple would help to ensure Office's value was available to everyone, and Apple was committing to make its iOS really show off the great things Office can do, which would further solidify Microsoft as the top developer for Apple.

On launch day, Apple's Senior Vice President for Worldwide Marketing, Phil Schiller, teased the audience as he set up the next demo at the iPad Pro launch. "We've been lucky to have some developers come in to work with us on professional productivity. Who knows productivity more than Microsoft?"

Nervous laughter filled the room.

"Yeah, these guys know productivity."

Kirk Koenigsbauer, the head of Office marketing, took the stage to proclaim that more than ever we are doing great work for the iPad.

But the publicity value of working with old rivals was far down on my list of motivations for pursuing them. Sure, people like to hear about competitors getting along. But forging great business partnerships is too difficult if PR is the sole purpose. For me, partnerships—particularly with competitors—have to be about strengthening a company's core businesses, which ultimately centers on creating additional value for the customer. For a platform company, that means doing new things with competitors that can accrue value back to one of the platforms.

Sometimes that means working with old rivals and sometimes it means forging surprising new partnerships. We work with Google, for example, to make it possible for Office to work on their Android platform. We partner with Facebook to make all of their applications work universally across Windows products and, likewise, to help them make our Minecraft gaming applications work on their Oculus Rift, a virtual reality device that competes for attention with our own HoloLens. Similarly, we're working with Apple to enable customers to better manage their iPhones within an enterprise. And we're working with Red Hat, a Linux platform that competes with Windows, so that enterprises built on Red Hat can use our Azure cloud to scale up globally by taking advantage of investments we've made in local data centers around the world. Our partnership with Red Hat may not be as surprising to some as our work with Apple and Google, but when I stood onstage with a slide just over my shoulder proclaiming "Microsoft ♥ Linux," one analyst concluded that hell must have frozen over.

Partnerships like these can exist, at times uneasily, with competitors in specific product or service categories. We compete vigorously with Amazon in the cloud market; there's no ambiguity about that. But why can't Microsoft and Amazon partner in other areas? For example, Bing powers the search experience on Amazon Fire tablets.

We have to face reality. When we have a great product like Bing, Office, or Cortana but someone else has created a strong market position with their service or device, we can't just sit on the sidelines. We have to find smart ways to partner so that our products can become available on each others' popular platforms.

In today's era of digital transformation, every organization and every industry are potential partners. Consider the taxi and entertainment industries. Ninety percent of Uber riders wait less than ten minutes for a driver, compared with 37 percent of taxi riders. Netflix costs its viewers $0.21 per hour of entertainment compared with $1.61 per hour with the old Blockbuster video-rental model. These are some of the higher visibility examples of digital transformation, but it's happening in every industry. We estimate the value of these transformations over the coming decade to be about $2 trillion.

Companies are focused on ensuring that they stay relevant and competitive by embracing this transformation. And we want Microsoft to be their partner. To do so, there are four initiatives every company must make a priority. The first is engaging their customer base by leveraging data to improve the customer experience. Second, they must empower their own employees by enabling greater and more mobile productivity and collaboration in the new digital world of work. Third, they must optimize operations, automating and simplifying business processes across sales, operations, and finance. Fourth, they must transform their products, services, and business models.

Every company is becoming a digital company, and that process begins with infusing their products with intelligence. Experts estimate between 20–50 billion "connected things" will be in use by 2020, presenting a significant opportunity for companies to drive their own digital transformation. GE has become a full-blown digital company with its Predix platform, which partners with Microsoft to connect industrial equipment, analyze data from those machines, and deliver real-time insights. Toyota has a connected auto division that has transformed their cars and trucks into next-generation digital-era vehicles—moving digital platforms that enable cars to communicate with other cars and even with the city's infrastructure. Rolls-Royce is designing their engines as big-data platforms to predict failures and minimize breakdowns.

Our emphasis on strategic partnerships isn't really new. It's actually another example of how we have been rediscovering the soul of Microsoft. When I look at our founders, Paul Allen clearly saw the power of new computers and Bill Gates saw the power of software. Together they were able to create magic and, more importantly, democratize computing. I sometimes wonder: If Bill and Paul had not succeeded with Microsoft, what would the world look like? Would we have independent hardware manufacturers, independent software vendors, system integrators, and others? Our original business model was built on an ecosystem of partners—independent software developers like Adobe and Autodesk, video game makers like EA Sports, hardware manufacturers like Dell, HP, and Lenovo, and retailers like Best Buy. I don't think Google would have existed but for the PC browser. Microsoft enabled Google to build a toolbar for our Internet Explorer, making Google services more visible and accessible. As a result of these and other partnerships, Microsoft and the PC helped nurture a host of billion-dollar companies—and Microsoft attracted millions of additional customers in the process.

When I became CEO, I sensed we had forgotten how our talent for partnerships was a key to what made us great. It's the kind of thing that can happen to any great company. Success can cause people to unlearn the habits that made them successful in the first place. We knew we needed to retrain our partnership muscles. We had to look anew at our industry and find ways to add value for our customers whether they were on an Apple device, a Linux platform, or an Adobe product.

Fortunately, this instinct is part of my DNA. My very first job at Microsoft in 1992 was all about partnering. We were building Windows NT, a 32-bit operating system. But most of the back-end applications that we needed to become viable had been built for Unix-based minicomputers, not Windows. And so my task as a young Windows NT technical evangelist was to move those applications onto the PC architecture. Lacking credibility as a serious enterprise player, Microsoft had to do a lot of hard work just to be considered. We built prototypes of applications for our PC platform, and then took them to customers in manufacturing, retail, and health care to show them that their big, robust minicomputer apps really could run just as well on a PC—maybe even better. They were surprised to see their mission-critical apps work with a graphical user interface on a device they'd thought of as a toy.

I can distinctly remember one of our first design wins. In retail, point-of-sales devices are ubiquitous and a lucrative market for technology. But there was no software standard to ensure that the cash register, the scanner, and other retail peripherals would all work together with the backend accounting and inventory systems. So my colleagues and I wrote up the standards and the specs that made it possible for Windows to enter the point-of-sales market. We started with nothing, but built up a major enterprise business.

To be sure, partnering has its challenges, even with long-standing partners. Sometimes we have to revive old relationships. Take Dell, for example, which over the years has shipped hundreds of millions of Windows computers. In 2012, when Microsoft decided for the first time to design and produce its own line of hardware, the Surface series, we morphed from pure partner into something murkier—a partner and a direct competitor. Then, to make things even more ambiguous, Dell took aim at one of Microsoft's cherished businesses by purchasing EMC, a leading producer of cloud technologies. It remains one of the largest technology acquisitions in history. Yet, through it all, Dell and Microsoft continued to partner in areas of mutual benefit—such as Dell licensing Windows for its laptops and selling Microsoft Surface products through its massive global distribution operation. In fact, Dell, HP, and others saw the popularity of Surface and began to innovate with their own new line of two-in-one computers.

But the trade press wondered if the lifelong partnership between our two companies was on the ropes. Just after becoming CEO, I joined Michael Dell in Austin, Texas, at his annual strategy day dedicated to answering questions from the press and stock analysts. In 2015, just after the EMC merger, a puzzled Emily Chang of Bloomberg News asked Michael and me to describe our relationship: "Are you friends? Are you frenemies?"

It's a simple question, and I offered a simple answer: "We are longtime friends who compete for and serve many of the same customers." But the real answer requires a more in-depth description.

In the 1990s, Microsoft developed a reputation for being a tough partner, to put it mildly. Documents and testimony in the U.S. Department of Justice's antitrust case against Microsoft (not to mention news stories and books) were filled with often damning stories of a company moving fast, competing hard, and upsetting many a partner. The government took action, the competitive landscape shifted, and now our mission and culture are different. A company that once was seen as crushing the competition is now focused on achieving business growth by empowering everyone on the planet.

I was part of the hard-driving Microsoft of the 1990s, but I wasn't personally engaged in the antitrust case. In fact, back then I was begging for customers and partners to work with us on our fledgling server business, a job that demanded an attitude of humility rather than one of hubris. One lesson I learned from the antitrust case (there were many lessons) was to compete hard and then equally celebrate the opportunities we create for everyone. It's not a zero-sum game.

I've taken that to heart. Google today is a dominant company in our industry. For years we've competed in the marketplace while also feuding through nonstop complaints to government regulators in the United States and abroad. As CEO, I decided to turn the page on that strategy, reasoning that it was time to end our regulatory battles and focus all of our energy on competing for customers in the cloud. Sundar Pichai, Google's CEO, is a competitor who I also count as a friend. After a series of very productive discussions and thoughtful negotiations between our two organizations, led by Brad Smith, Microsoft's president and chief legal officer, Sundar and I surprised observers of the rivalry with a joint statement: "Our companies compete vigorously, but we want to do so on the merits of our products, not in legal proceedings."

In pushing this change of attitude, I've been helped by the simple fact that I am a fresh face, new blood. Losing the baggage of history makes it easier for me to break down old walls of mistrust. But will it be enough?

—

Early in my tenure as CEO, I decided I needed to talk with Peggy Johnson, who had been doing an amazing job at managing partnerships and business development at Qualcomm, a semiconductor and wireless telecom company based in San Diego. I called her up one Saturday afternoon at her home in the San Diego area and asked her whether she would consider joining Microsoft. I could tell that she was skeptical and even felt a little disloyal having the conversation. I managed to persuade Peggy to meet me for dinner in Silicon Valley.

Walking into the Four Seasons Hotel for our meeting, a few people recognized me and heads turned with more than a little curiosity. We were seated at a quiet table and soon found ourselves talking excitedly about ambient intelligence—the notion that more and more objects in our homes, offices, and other spaces will automatically recognize our human presence and respond to our preferences. For Microsoft to successfully lead that digital transformation would take new, untraditional, surprising partnerships, investments, and acquisitions. I sensed that Peggy was hooked on that vision too. Later I learned that she'd called her husband immediately after our meal to persuade him that Redmond, Washington, was in their future. She had the job and the direction to help "make Silicon Valley our best friend."

Peggy's ease, humility, and passion for technology impressed me. These were just the qualities I wanted Microsoft to convey to our potential business partners. Little did I know how soon those attributes would be called into action.

One of our major partnership goals was to build Microsoft applications for competing platforms like Google's Android operating system and for Apple's iOS. We needed to have our apps preloaded on phones with varying operating systems so that when a consumer bought a phone, Microsoft apps would already be there.

One of the important partners we needed to work with on this front was Samsung, the Korean manufacturer of the world's most popular Android smartphone. We'd partnered with Samsung for more than thirty years. But in the summer of 2014, as Peggy was preparing to move to Redmond, Microsoft's relationship with Samsung was breaking down. Several years earlier, Samsung had entered into an agreement to license some of Microsoft's intellectual property, but since then the company's smartphone sales had quadrupled and its Android phones were now the bestselling in the world. After Microsoft announced it would purchase the devices and services division of Nokia, the Finnish smartphone manufacturer, Samsung informed us they would no longer abide by the contract we had signed. Samsung's president, Jong-Kyun (J.K.) Shin, had become so upset that he refused to meet with anyone from Microsoft. The partnership was sharply tilting toward litigation.

We worked to get Peggy up to speed on the Samsung relationship. She read documents from both sides, asked good questions, and offered creative ideas about how we might resolve our differences. Fortunately for us, Peggy had developed a great relationship with J.K. He agreed to meet with her. She and a team from business development and legal journeyed to his office in Seoul, where they found the room filled with people known for their tough negotiation style. Peggy and the team worked to demonstrate respect throughout the meeting. Rather than make demands, she decided instead to listen, reserving judgment and seeking to empathize with Samsung's perspective.

She and the team returned to Redmond not as advocates for one side, but motivated to find a solution in the middle. Though new to the company, Peggy already exemplified the culture we aspired to have. She saw what was possible by keeping a growth mindset, not one fixed on pointing fingers or assigning blame. She and her team brought everyone to the table, demonstrating what diversity and inclusion looks like. And she showed us the importance of getting out of our headquarters at Redmond— away from our insular, comfortable world—and inside that of our partners and customers.

In the end, we still had to resolve some of our issues through the courts, but we also continued to show respect. "Microsoft values and respects our partnership," we wrote in a statement. "Unfortunately, even partners sometimes disagree."

Today Microsoft apps are popular on Samsung smartphones; Windows 10 powers Samsung tablets and its ambitions for the far-flung Internet of Things.

Around the same time, we were embroiled in a contentious dispute with Yahoo, which used the Bing search engine as its exclusive search partner. Microsoft and Yahoo shared in the revenue from the searches performed by Bing. But, as with Samsung, our relationship with Yahoo was deteriorating as Yahoo's business model came under pressure, and lawsuits were being threatened. Yahoo wanted to breach its contract.

We worked to mend the relationship not by presenting a list of demands, but by listening, empathizing with a partner's situation and exploring ideas. In the end, we decided to forgo the requirement of exclusivity for Bing as Yahoo's search partner. The issue was generating too much needless friction between the two parties, and we were confident that our technology and our partnership would prevail. We avoided costly litigation, and today Bing continues to handle the majority of Yahoo's searches.

These experiences have taught us a lot and refreshed our partnering spirit. Microsoft already has the largest ecosystem of partners in the world. Hundreds of thousands of companies worldwide build and sell solutions that support our products and services. In addition, millions of customers in every sector have built their businesses and organizations using Microsoft technologies. My ultimate goal is to be the biggest platform provider underneath all of this entrepreneurial energy, with an unrelenting focus on creating economic opportunity for others.

But if we want to convince millions of new companies around the world to bet on our platform, we need to start by earning their trust. In Chapter 7, I will explicitly explore the notion that trust is built by being consistent over time. It's built by being clear that there are places where we are going to compete to be best in class, and there are places where we can work together to add value for each other's customers.

Trust has many other components as well—respect, listening, transparency, staying focused, and being willing to hit reset when necessary. We have got to be principled about it.

Partnerships are journeys of mutual exploration, and so we need to be open to unexpected synergies and fresh ways to collaborate. Openness begins with respect—respect for the people at the table and the experiences they bring, respect for the other company and its mission. Do we always agree? Of course not. But we always seek to listen intelligently, seeking to understand not just the words we are hearing but the underlying intentions. I try hard not to bring needless history into the room, and I don't let the limitations of the past dictate the contours of the future.

Over the years, I've found that openness is the best way to get things done and to ensure all parties feel terrific about the outcome. In a world where innovation is continuous and rapid, no one has time to waste on unnecessary cycles of work and effort. Being straightforward with one another is the best way to achieve a mutually agreeable outcome in the fastest time possible.

When complications threaten to stymie the effort to build a partnership, it helps to stay focused on long-term goals. Rather than being distracted by the endless opportunities to collaborate and the numerous questions they raise, I like to start with one or two areas of focus. Once companies can work together successfully, then they can tackle the next set of ideas and challenges.

Finally, don't be afraid to take a pause. Even when both parties have nothing but the best intentions, things can sometimes go sideways and may even come to a standstill. Sometimes it's critical to look at an existing relationship with a fresh set of eyes. A strategy that failed in the past might work in the future. Technology changes. The business environment changes. People change. It's a mistake to write off any relationship as a lost cause. Tomorrow always begins with a chance to create new opportunities.

This approach led to real breakthroughs in our partnership with a standard-bearer in the creative world—Adobe, a pioneer in font development and the maker of PhotoShop, Illustrator, Acrobat, Flash, and many other products loved by artists and designers. Adobe was built on Windows, but we competed on document standards and, over the years, in spite of many common customers we simply drifted apart. My friend from Hyderabad Public School, Shantanu Narayen, had become CEO of Adobe earlier and when I was named CEO of Microsoft our two companies began to reengage. We still compete in areas of overlap, but we have a much deeper partnership in which Adobe's creative software is now the inspiration for new Microsoft devices like

Surface Studio and Surface Hub. Together we are transforming what artists can do with a computer. And we've expanded beyond the creative cloud into Adobe's marketing cloud, which is built on top of our Azure platform.

I am often asked, "When is a partnership appropriate as opposed to an acquisition?" The answer is best framed as another question, "Can we create more value for customers by coming together as one entity or as two?" In my experience, whether we're talking about a gigantic acquisition like our deal to purchase the social network LinkedIn or smaller acquisitions like those of app developers Xamarin, Acompli, and MileIQ, the acquisitions that succeed generally start as partnerships born out of careful analysis of customer needs. That was the case with LinkedIn, which Microsoft acquired in 2016 for $26 billion, one of the largest such deals in history. For more than six years, Microsoft and LinkedIn worked together to enable our one billion users and their nearly half-a-billion members—the Venn diagram of our customers would overlap 100 percent with theirs—to synchronize contacts so that Office contacts were available in LinkedIn and vice versa. Microsoft made its technical specifications available so that LinkedIn could build a beautiful app for Windows and partnered on a Social Connector that enabled rich connections and collaboration across both platforms. For us to do even further integration, and create more compelling scenarios and value for our customers, we had to come together as one.

Together we built not just a track record, but also a shared vision and mutual trust. That's why on the day we announced the acquisition, LinkedIn's CEO Jeff Weiner explained the deal to technology reporter Kara Swisher by saying, "You look at how Microsoft is increasingly becoming more agile, more innovative, more open, more purpose driven. And that played a big role in this."

Going back to my days as an engineer, I've used the following mental model to capture how I manage time:

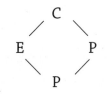

Employees. Customers. Products. Partners. Each element needs time, attention, and focus if I'm going to create the value for which I am ultimately accountable. All four are important, and without discipline even the best managers can overlook one or more. Employees and products command attention every day, as they are closest to us; customers provide the resources we need to do anything, so they also command energy. But partners provide the lift we need to soar. They help us see around corners, help us locate new opportunities we might not see alone. Since becoming a CEO I now recognize there are many more constituents in this constellation. Governments and communities, for example, are critical, too. There has to be a disciplined approach in which all of these players see the value of a company, of its products and services. Maximizing value comes from maximizing the well-being and vibrancy of all these constituents.

CHAPTER 6

Beyond the Cloud

Three Shifts: Mixed Reality, Artificial Intelligence, and Quantum Computing

Originally, I thought of this book as a collection of meditations from a CEO in the midst of transformation. As someone both navigating a corporate transformation and creating transformational technologies, my aim was to share these experiences in real time rather than look back on them years later. The Microsoft transformation, of course, is ongoing. In the face of global economic and technological uncertainty, we reset our mission, reprioritized our culture, and built or rebuilt strategic partnerships in order to solidify the foundation of our business. We also needed to hasten our innovative spirit and place new, bold bets. This is what has made Microsoft a trusted tech brand for more than forty years.

We looked beyond the PC and the server to drive success in

the cloud. But we also had to look beyond the cloud. Forecasting technology trends can be perilous. It's been said we tend to overestimate what we can achieve in the short run, but underestimate what can be achieved in the long run. But we are investing to lead in three key technologies that will shape our industry and others in the years to come—mixed reality, artificial intelligence, and quantum computing. These technologies will inevitably lead to massive shifts in our economy and society. In the final three chapters of this book, I'll explore the values, ethics, policies, and economics we need to consider in preparation for this next wave.

Here is one way to think about the convergence of these coming technology shifts. With mixed reality we are building the ultimate computing experience, one in which your field of view becomes a computing surface and the digital world and your physical world become one. The data, apps, and even the colleagues and friends you think of as being on your phone or tablet are now available anywhere you want to access them—while you're working in your office, visiting a customer, or collaborating with colleagues in a conference room. Artificial intelligence powers every experience, augmenting human capability with insights and predictive power that would be impossible to achieve on our own. Finally, quantum computing will allow us to go beyond the bounds of Moore's Law—the observation that the number of transistors in a computer chip doubles roughly every two years—by changing the very physics of computing as we know it today, providing the computational power to solve the world's biggest and most complex problems. MR, AI, and quantum may be independent threads today, but they are going to come together. We're betting on it.

A technology company that misses multiple trends like these will inevitably fall behind. At the same time, of course, it's dangerous to chase untested future technologies while neglecting the core of the current business. That's the classic innovator's dilemma—to risk existing success while pursuing new opportunities.

Historically, Microsoft has struggled at times to get this balance right. We actually had a tablet before the iPad; we were well along the path toward an e-reader before the Kindle. But in some cases our software was ahead of the key components required for success, such as touchscreen hardware or broadband connectivity. In other cases, we lacked end-to-end design thinking to bring a complete solution to market. We also got a bit overconfident in our ability to fast-follow a competitor, forgetting that there is inherent risk in such a strategy. We were perhaps timid in disrupting our own highly successful business models. We've learned from all this. There is no formula to inventing the future. A company has to have a complete vision for what it can uniquely do, and then back it up with conviction and the capability to make it happen.

I had decided before becoming CEO that we would need to continue to invest, and to do so in an aggressive and more focused way, in new technologies and new markets—but only if we could satisfactorily meet our three Cs—do we have an exciting *concept,* do we have the *capabilities* necessary to succeed, and a *culture* that welcomes these new ideas and approaches?

To avoid being trapped by the innovator's dilemma—and to move from always focusing on the urgency of today to considering the important things for tomorrow—we decided to look at our investment strategy across three growth horizons: first, grow today's core businesses and technologies; second, incubate new ideas and products for the future; and third, invest in long-term breakthroughs. On horizon one, our customers and partners will continue to see quarter-by-quarter, year-by-year innovations in all of our businesses. On horizon two, we're already investing in some exciting nearer-term platform shifts, such as new user interfaces with speech or digital ink, new applications with personal assistants and bots, and Internet of Things experiences for everything from factories to cars to home appliances. On horizon three, Microsoft is highly focused in areas that only a few years ago sounded distant, but today are frontiers of innovation—mixed reality, artificial intelligence, and quantum computing. Mixed reality will become an essential tool in medicine, education, and manufacturing. AI will help forecast crises like the Zika epidemic and help us focus our time and attention on things that matter most. Quantum computing will give us the computational power to cure cancer and effectively address global warming.

The intellectual history of how computers augment the human intellect and build a collective IQ has always fascinated me. Doug Engelbart in the 1960s performed "the mother of all demos," introducing the mouse, hypertext, and shared-screen teleconferencing. Engelbart's Law states that the rate of human performance is exponential; that while technology will augment our capabilities, our ability to improve upon improvements is a uniquely human endeavor. He essentially founded the field of human-computer interaction. There are many other visionaries who influenced me and the industry, but around the time I joined Microsoft in 1992, two futuristic novels were being eagerly consumed by engineers all over campus. Neal Stephenson's *Snow Crash* popularized the term *metaverse*, envisioning a collective virtual and shared space. David Gelernter wrote *Mirror Worlds*, foreseeing software that would revolutionize computing and transform society by replacing reality with a digital imitation. These ideas are now within sight.

—

It is a magical feeling, at least for me, the first time you experience a profound new technology. In the 1980s, when I first learned to write a few lines of BASIC code for that Z80 computer my dad bought for me, the lightbulb went off. Suddenly I was communicating with a machine. I wrote something and it generated output, a response. I could change the program and instantly change the response. I had discovered software, the most malleable resource humans have made. It was like lightning in a bottle. I clearly remember the excitement I felt the first time I encountered the spreadsheet. A data structure like pivot tables was now second nature to how one thought about numbers.

Our industry is full of those eureka moments of discovery. My most startling moment arrived, surprisingly, on the surface of planet Mars—standing in the basement of Microsoft's Building 92.

It was there that I first slipped on a HoloLens device, a small head-mounted computer that is completely self-contained. Suddenly HoloLens transported me—virtually, of course—onto the surface of the Red Planet, 250 million miles away, thanks to a feed from NASA's Mars rover, *Curiosity*. Through HoloLens, I could see my two street shoes walking, in the most convincing and baffling way, on the dusty Martian plain near a rocky waypoint called Kimberley along the rover's journey to Murray Buttes. HoloLens made it possible for me both to walk around the actual room—to see a desk and to interact with people around me—and to inspect rocks on Mars's surface. That's the amazing, unprecedented nature of what we call mixed reality. The experience was so inspiring, so moving, that one member of my leadership team cried during that virtual excursion.

What I saw and experienced that day was a glimpse of Microsoft's future. Perhaps this particular moment will be remembered as the advent of a mixed reality revolution, one in which everyone works and plays in an immersive environment that blends the real world and a virtual world. Will there one day be mixed reality natives—young people who expect all of their computer experiences to be immersive blends of the real and the virtual—just as today we recognize digital natives, those for whom the Internet has always been there?

Companies are taking different approaches with head-mounted computers. *Virtual reality*, as provided by our Windows 10 MR devices or Facebook's Oculus Rift, largely blocks out the real world, immersing the user in a completely digital world. Google Glass, for example, projects information onto your eyeglasses. Snapchat Spectacles let you augment what you see with relevant content and filters. HoloLens provides access to *mixed reality* in which the users can navigate both their current location—interact with people in the same room—and a remote environment while also manipulating holograms and other digital objects. Analysts at Gartner Inc., the technology research firm, have made an art from the study of the hype cycles and arcs followed by new technologies as they move from invention to widespread adoption (or demise), and believe virtual reality technologies are likely five to ten years away from mainstream adoption.

Just getting to the starting line proved difficult for us. My colleague Alex Kipman had been perfecting a prototype of HoloLens for some time. Alex and his team had already created one breakthrough: They'd developed Microsoft Kinect, the motion-sensing technology that today is an ingredient in leading-edge robots (enabling them to move in a more human-like manner), while also providing a fun way of using your body to play games on Xbox. However, Alex's HoloLens project had bounced around the company in search of continued funding. It was unclear whether Microsoft would invest in mixed reality, a new business in an unproven market. The quest seemed so ridiculous at times that Alex whimsically code-named the project Baraboo in honor of a town in Wisconsin that is home to a circus and clown museum.

Once I got a chance to see what HoloLens could do, I was sold. While HoloLens has obvious applications in video gaming, I instantly saw its potential in classrooms, hospitals, and, yes, space exploration. NASA was, in fact, one of the first organizations to see the value of HoloLens, adopting an early version to enable astronauts on Earth to collaborate with astronauts in space. If anyone was on the fence after the Mars demonstration, Bill Gates's email after his experience convinced even the most skeptical.

I was VERY impressed with 2 things about the Mars demo: First, the fidelity was VERY good. The image looks real and when I moved my head it felt real. I felt like I was there. Second, the ability to move physically around the space was quite natural while using peripheral vision to avoid hitting anything. Although I am still not sure what applications will take off, the latest demo really has me enthused about the project and that we will find a way to make this a success. I have been converted.

Yes, Alex, we'll invest.

To understand the soul of HoloLens, it helps to understand Alex and his past. In some ways, our stories have a lot in common. The son of a career government diplomat in Brazil, Alex moved around a lot as a kid and found that math, science, and eventually computers were his only consistent companions. "If you know how to paint with math and science, you can make anything," he once told me. His parents bought him an Atari 2600 home video console that he broke repeatedly but eventually learned to program. His passion for technology led him to the Rochester Institute of Technology, an internship with NASA, and, later, highly sophisticated computer programming roles in Silicon Valley.

His quest, however, was to find a place where he could design software for the sake of software, a place that treated software as an art form. He came to Microsoft where he would play a role in designing Windows Vista, the long-awaited successor to Windows XP. When Vista received lukewarm reviews despite its advanced features, no one was more disappointed than Alex. He took it personally and returned to Brazil to reflect—to hit refresh on his own career outlook. Alex is very philosophical and turned to Nietzsche for direction: "He who has a 'why' to live for can bear almost any 'how.'" Alex was upset with himself because he did not yet have his "why," a point of view about where computing should be headed.

He would later tell journalist Kevin Dupzyk that he visited a farm on Brazil's eastern shore, wandering around with a notebook and pondering the contribution he wanted to make to computing. He began to think about how computing could displace time and space. Why are we chained to keyboards and screens? Why can't I use my computer to be with anyone I want, no matter where they are? Alex sensed that the evolution of computing had only reached the equivalent of prehistoric cave paintings. MR was to become a new paintbrush that would create an entirely new computing paradigm.

Alex defined a new career quest for himself: "I am going to make machines that perceive the real world." Perception—not a mouse, keyboard, and screen—would be the protagonist of his story. Machines that perceive us became his "why."

The "how," the blueprint, became to build a new computing experience designed around sensors that can perceive humans, their environment, and the objects around them. This new computing experience must enable three kinds of interactions: the ability to input analog data, the ability to output digital data, and the ability to feel or touch data—something known as *haptics*.

Kinect was the first step in this journey—it checked the box for a human to provide input to a computer simply by moving. Suddenly we could dance with a computer. Now HoloLens is checking multiple boxes. It enables humans, the environment, and objects to give and receive input and output across time and space. Suddenly an astronaut on Earth can inspect a crater on Mars. The final piece, the haptics, will include the ability to touch and feel. When we dance using Kinect or reach for a rock using HoloLens, we cannot yet feel our dance partner or that rock. But one day we will.

Today, our focus at Microsoft is to democratize mixed reality, to make it available to everyone. Our launch of HoloLens has been based on a proven strategy for Microsoft—inviting outside developers to help us create imaginative applications for the HoloLens platform. Soon after we announced HoloLens, more than five thousand developers submitted ideas for applications they wanted to build. We ran a twenty-four-hour Twitter poll to ask which idea we should build first. Developers and fans chose Galaxy Explorer, which enables you to look out your window and navigate the Milky Way—moving through it at your own pace, zooming in, annotating what you see, and storing the experience for later. It replicates the environment of a planet on your room's walls—dusty winds, hot plasma, and ice formations.

Now other developers are crafting tremendously useful new

applications for HoloLens. Lowe's home improvement stores, for example, are using HoloLens to allow their customers to stand in their own kitchens and bathrooms, and superimpose holograms of new cabinets, appliances, and accessories so they can see exactly what their remodel will look like.

The trajectory of this technology begins with simply tracking what the machine is seeing but someday will completely understand more complex tasks, which we'll learn about as we get to artificial intelligence. Kinect gave a computer the ability to track your movements—to see you and make sense of your actions. That's where AI, machine learning, and mixed reality are today. Technology can increasingly see, speak, and analyze, but it cannot yet feel. But mixed reality may help machines empathize with humans. Through these technologies, we will be able increasingly to experience what a refugee or a crime victim experiences, potentially enhancing our ability to make emotional connections across barriers that currently divide people from one another. In fact, I had a chance to meet several student developers from Australia who participated in our Imagine Cup competition. They built an MR application that helps certain caregivers learn to see the world through the eyes of someone with autism.

———

Artificial Intelligence has been portrayed in myriads of ways by Hollywood, which has practically made the technology its own subgenre. In 1973's *Westworld*, Yul Brenner plays a robot—an AI-infused, tough-guy cowboy—who walks into a saloon itching for a gunfight. Year's later, Disney had a different depiction. In its *Big Hero 6*, a pillowy giant robot named Baymax lovingly helps his fourteen-year-old owner get through a suspenseful journey. "He'll change your world," the film proclaims.

And that's just it. AI will change our world. It will augment and assist humans, much more like Baymax than Brenner.

A confluence of three breakthroughs—Big Data, massive computing power, and sophisticated algorithms—is accelerating AI from sci-fi to reality. At astonishing rates, data is being gathered and made available thanks to the exponential growth of cameras and sensors in our everyday life. AI needs data to learn. The cloud has made tremendous computing power available to everyone, and complex algorithms can now be written to discern insights and intelligence from the mountains of data.

But far from Baymax or Brenner, AI today is some ways away from becoming what's known as artificial general intelligence (AGI), the point at which a computer matches or even surpasses human intellectual capabilities. Like human intelligence, artificial intelligence can be categorized by layer. The bottom layer is simple pattern recognition. The middle layer is perception, sensing more and more complex scenes. It's estimated that 99 percent of human perception is through speech and vision. Finally, the highest level of intelligence is cognition—deep understanding of human language.

These are the building blocks of AI, and for many years Microsoft has invested in advancing each of these tiers—statistical machine learning tools to make sense of data and recognize patterns; computers that can see, hear, and move, and even begin to learn and understand human language. Under the leadership of our chief speech scientist, Xuedong Huang, and his team, Microsoft set the accuracy record with a computer system that can transcribe the contents of a phone call more accurately than a human professional trained in transcription. On the computer vision and learning front, in late 2015 our AI group swept first prize across five challenges even though we only trained our system for one of those challenges. In the Common Objects in Context challenge, an AI system attempts to solve several visual recognition tasks. We trained our system to accomplish just the first one, simply to look at a photograph and label what it sees. Yet, through early forms of transfer learning, the neural network we built managed to learn and then accomplish the other tasks on its own. It not only could explain the photograph, but it was also able to draw a circle around every distinct object in the photograph and produce an English sentence that described the action it saw in the photo.

I believe that in ten years AI speech and visual recognition will be better than a human's. But just because a machine can see and hear doesn't mean it can truly learn and understand. Natural language understanding, the interaction between computers and humans, is the next frontier.

And so how will AI ever live up to its hype? How will AI scale to benefit everyone? Again, the answer is layered.

Bespoke. Today we are very much on the ground floor of AI. It is bespoke, customized. Tech companies with privileged access to data, computing power, and algorithms handcraft an AI product and make it available to the world. Only a few can make AI for the many. This is where most AI is today.

Democratized. The next level is democratization. As a platform company—one that has always built foundational technologies and tools upon which others can innovate—Microsoft's approach is to put the tools for building AI in the hands of everyone. Democratizing AI means enabling every person and every organization to dream about and create amazing AI solutions that serve their specific needs. It's analogous to the democratization that movable type and the printing press created. It's estimated that in the 1450s there were only about thirty thousand books in Europe—each one handcrafted by someone working in a monastery. The Gutenberg Bible was the first book produced using movable type technology, and within fifty years the number of books grew to an estimated 12 million, unleashing a renaissance in learning, science, and the arts.

That's the same trajectory we need for AI. To get there we have to be inclusive, democratic. And so our vision is to build tools that have true artificial intelligence infused across agents, applications, services, and infrastructure:

- We're harnessing artificial intelligence to fundamentally change how people interact with *agents* like Cortana, which will become more and more common in our lives.
- *Applications* like Office 365 and Dynamics 365 will have AI baked-in so that they can help us focus on things that matter the most and get more out of every moment.
- We'll make the underlying intelligence capabilities of our own services—the pattern recognition, perception, and cognitive capabilities—available to every application developer in the world.
- And, lastly, we're building the world's most powerful AI super-computer and making that *infrastructure* available to anyone.

A range of industries are using these AI tools. McDonald's is creating an AI system that can help its workers take your order in the drive-through line, making ordering food simpler, more efficient, and more accurate. Uber is using our cognitive services tools to prevent fraud and improve passenger safety by matching the driver's photograph to ensure the right driver is at the wheel. And Volvo is using our AI tools to help recognize when drivers are distracted to warn them and prevent accidents.

If you're a business owner or manager, imagine if you had an AI system that could literally see your entire operation, understand what's happening, and notify you about the things you care most about. Prism Skylabs has innovated on top of our cognitive services so that computers monitor video surveillance cameras and analyze what's happening. If you have a construction company, the system will notify you when it sees the cement truck arrive at one of your work sites. For retailers, it can keep track of inventory or help you find a manager in one of your stores. One day, in a hospital setting, it might watch the surgeon and supporting staff to warn the team, before it's too late, if it sees a medical error.

Learn to Learn. Ultimately, the state of the art is when computers learn to learn—when computers generate their own programs. Like humans, computers will go beyond mimicking what people do and will invent new, better solutions to problems. Deep neural networks and transfer learning are leading to breakthroughs today, but AI is like a ladder and we are just on the first step of that ladder. At the top of the ladder is artificial general intelligence and complete machine understanding of human language. It's when a computer exhibits intelligence that is equal to or indistinguishable from a human.

One of our top AI researchers decided to try an experiment to demonstrate how a computer can learn to learn. A highly esteemed computer scientist and medical doctor, Eric Horvitz, runs our Redmond research lab and has long been fascinated with machines that perceive, learn, and reason. His experiment was to make it easier for a visitor to find him, and to free up his human assistant for more critical work than the mundane task of constantly giving directions. So, to visit his office, you enter the ground floor lobby where a camera and computer immediately notices you, calculates your direction, pace, and distance and then makes a prediction so that an elevator is suddenly waiting for you. Getting off the elevator, a robot says hello and asks if you need help finding Eric's desk among the confusing corridors and warren of surrounding offices. Once there, a virtual assistant has already anticipated your arrival, knows Eric is finishing up a phone call, and asks if you'd like to be seated until Eric is available. The system received some basic training but, over time, learned to learn on its own so that programmers were not needed. It was trained, for example, to know what to do if someone in the lobby pauses to answer a call or stops to pick up a pen that's fallen on the floor. It begins to infer, to learn, and to program itself.

Peter Lee is another gifted AI researcher and thinker at Microsoft. In a meeting one morning in his office, Peter reflected on something the journalist Geoffrey Willans once said. "You can never understand one language until you understand at least two." Goethe went further. "He who does not know foreign languages does not know anything about his own." Learning or improvement in one skill or mental function can positively influence another one. The effect is transfer learning, and it's seen not only in human intelligence but also machine intelligence. Our team, for example, found that if we trained a computer to speak English, learning Spanish or another language became faster.

Peter's team decided to invent a real-time, language-to-language translator that breaks the language barrier by enabling a hundred people at one time to speak in nine different languages, or type messages to one another in fifty different languages. The result is inspiring. Workers all over the globe can be linked via Skype or simply by speaking into their smartphones and understand one another instantly. A Chinese speaker can present a sales and marketing plan in her native language and teammates listening in can see or hear what's being said in their native languages.

My colleague, Steve Clayton, told me the story of how profound this technology was for his multicultural family. He said the first time he saw the technology demonstrated he knew that his young children, English speakers, would for the first time be able to have a live conversation with their Chinese-speaking relatives.

Looking ahead, many others will use our tools to expand the translator beyond the initial languages we began with. A healthcare company, for example, may want to create English, Spanish, and other highly specialized versions of the translator that speak the language of medicine. An AI tool would be used to watch and listen to health-care professionals talk, and then, after a period of observation, would automatically generate a new model for a health-care–specific version. A Native American tribe might preserve its language by listening to elders speak. The optimal state will be when those AI systems not only translate but improve— perhaps converting conversation into ideas about improving patient care or converting a conversation into an essay.

The holy grail for AI has long been a really good personal agent that can assist you in meaningful ways to get the most out of life at home and work. Cortana, named for our synthetic intelligence character in the popular video game Halo, is a fascinating case study of where we stand today and how we hope one day to deliver a highly effective alter ego—an agent that knows you deeply. It will know your context, your family, your work. It will also know the world. It will be unbounded. And it will get smarter the more it's used. It will learn from its interactions with all of your apps as well as from your documents and emails in Office.

Today, there are more than 145 million Cortana users each month in 116 countries. Those customers already have asked 13 billion questions, and with each question the agent is learning to become more and more helpful. In fact, I've come to rely on Cortana's commitment feature, which searches through my emails hunting for promises I've made and then gently reminds me as the deadline approaches. If I told someone I'd follow up with them in three weeks, Cortana makes a note of that and reminds me later to ensure I keep my commitment.

Our Cortana team, part of a relatively new AI and research division, works in a tall Microsoft building in downtown Bellevue with windows looking out over the Pacific Northwest's lakes and mountains. The beauty of these surroundings coupled with the mandate to push the edges of innovation has attracted incredible talent—designers, linguists, knowledge engineers, and computer scientists.

Jon Hamaker, one of the group's engineering managers, says his goal is for customers to tell him "I couldn't live without Cortana—she saved me again today." He and his team spend their days thinking through scenarios that would make that true. What do our users do—how, when, where, and with whom do they interact? What would build a bond with the user? How can we save the user time, reduce the user's stress, help the user stay one step ahead of everyday challenges? Hamaker's quest is to capture every type of data from sources including GPS, email, calendar, and correlative data from the web and turn that data into understanding, and even empathy. Perhaps your digital assistant will schedule time to ask you questions that will help fill in gaps where the data is insufficient in order to help you more. Perhaps the assistant will be helpful in times of uncertainty— when you're in a new place where the currency and the language are foreign, for example.

Those kinds of uncertainties fascinate our engineers who focus on semantic ontologies, the study of interrelationships among people and entities. Their ambition is to develop an agent that can do much more than simply get you a search result. They dream of a day when a digital agent will understand context and meaning, using them to better predict what you need and want. The digital assistant should always have a good answer, sometimes even an answer to a question you didn't know you had.

Emma Williams is not an engineer. She was an English literature scholar with a focus on Anglo-Saxon and Norse literature. Her job is to think through the emotional intelligence (EQ) design of our AI products, including Cortana. She's confident about the IQ of the team we have working on agents; she wants to make sure we have the EQ as well.

One day she discovered a new build of Cortana in which Cortana displayed anger when asked certain questions. Williams promptly put her foot down. (If medieval Norse tales about Vikings taught her anything, it's that pillaging while searching for resources should not be part of discovering new things.) She made the point that Cortana offers an implicit promise to users that she will always be calm, cool, and collected. Rather than becoming angry with you, Cortana should understand your emotional state, whatever it is, and respond appropriately. The team revised Cortana in accordance with Williams's sensibilities.

If this journey toward an AI-powered assistant is one of a million miles, we've walked only the first few of those miles. But these first few steps are inspiring ones when we contemplate what they may produce.

My former colleague David Heckerman is a distinguished scientist who has spent thirty years working on AI. Years ago, he created one of the first effective spam filters by figuring out the weak link of his adversaries—the spammers who clog your in-box with junk mail—and foiled their attempts to succeed.

Today the team he built at Microsoft develops machine learning algorithms designed to discover and exploit weak links in HIV, the common cold, and cancer. HIV, the virus that causes AIDS, mutates rapidly and broadly in a human body, but there are constraints in how the virus mutates. The advanced machine learning algorithms we've built have discovered which sections of HIV proteins are absolutely essential to their function so that a vaccine can be trained to attack those very regions. Using clinical data, his team can simulate mutations and identify targets. Similarly, they are taking genomic sequences for a cancer tumor and predicting the best targets for the immune system to attack.

If the potential for this AI work is breathtaking, the potential for quantum computing is mind-blowing.

———

Santa Barbara, California, is closer to Hollywood than it is Silicon Valley. Its casual, beach-front college campus just north of Tinseltown is the unlikely center of quantum computing development, the very future of our industry. Its proximity to Hollywood is fitting, since a film script may be a better guide to quantum physics and mechanics than a textbook. Rod Serling's *The Twilight Zone* likely put it best: "You're traveling through another dimension, a dimension not only of sight and sound but of mind. A journey into a wondrous land whose boundaries are that of imagination. That's the signpost up ahead—your next stop, the Twilight Zone."

Defining quantum computing is no simple feat. Originating

in the 1980s, quantum computing leverages certain quantum physics properties of atoms or nuclei that allow them to work together as quantum bits, *or qubits*, to be the computer's processor and memory. By interacting with each other while being isolated from our environment, qubits can perform certain calculations exponentially faster than conventional, or classical, computers.

Photosynthesis, bird migration, and even human consciousness are studied as quantum processes. In today's classical computing world, our brain thinks and our thoughts are typed or spoken into a computer that in turn provides feedback on a screen. In a quantum world, some researchers speculate that there will be no barrier between our brains and computing. It's a long way off, but might consciousness one day merge with computation?

"If quantum *mechanics* hasn't profoundly shocked you, you haven't understood it yet," the Danish Nobel physicist Niels Bohr once said. A later Nobel physicist, Richard Feynman, proposed the notion of quantum computing, unleashing today's global pursuit to harness quantum mechanics for computation. Among those racing to understand it are Microsoft, Intel, Google, and IBM as well as startups like D-Wave and even governments with hefty national defense budgets. The shared hope is that quantum computing will utterly transform the physics of computing itself.

Of course, if building a quantum computer were easy, it would have been done by now. While classical computing is bound by its binary code and the laws of physics, quantum computing advances every kind of calculation—math, science, and engineering—from the linear world of bits to the multidimensional universe of qubits. Instead of being simply a 1 or a 0 like the classical bit, qubits can be every combination—a superposition—which enables many computations all at once. Thus, we enter a world in which many parallel computations can be simultaneously answered. In a properly constructed quantum algorithm, the result is, according to one of our scientists, "a great massacre in which all or most of the wrong answers are canceled out."

Quantum computing is not only faster than conventional computing, but its workload obeys a different scaling law—rendering Moore's Law little more than a quaint memory. Formulated by Intel founder Gordon Moore, Moore's Law observes that the number of transistors in a device's integrated circuit doubles approximately every two years. Some early supercomputers ran on around 13,000 transistors; the Xbox One in your living room contains 5 billion. But Intel in recent years has reported that the pace of advancement has slowed, creating tremendous demand for alternative ways to provide faster and faster processing to fuel the growth of AI. The short-term results are innovative accelerators like graphics-processing unit (GPU) farms, tensor-processing unit (TPU) chips, and field-programmable gate arrays (FPGAs) in the cloud. But the dream is a quantum computer.

Today we have an urgent need to solve problems that would tie up classical computers for centuries, but that could be solved by a quantum computer in a few minutes or hours. For example, the speed and accuracy with which quantum computing could break today's highest levels of encryption is mind-boggling. It would take a classical computer 1 billion years to break today's RSA-2048 encryption, but a quantum computer could crack it in about a hundred seconds, or less than two minutes. Fortunately, quantum computing will also revolutionize classical computing encryption, leading to ever more secure computing.

To get there we need three scientific and engineering breakthroughs. The math breakthrough we're working on is a topological qubit. The superconducting breakthrough we need is a fabrication process to yield thousands of topological qubits that are both highly reliable and stable. The computer science breakthrough we need is new computational methods for programming the quantum computer.

At Microsoft, our people and our partners right now are working with the transport, experimental and theoretical physics, and the mathematics and computer science that will one day make quantum computing a reality. The hotbed of this activity is Station Q, which is co-located with the theoretical physics department at the University of California at Santa Barbara. Station Q is the brainchild of Michael Freedman, who won math's top award, the Fields Medal, at the International Congress of the International Mathematical Union in 1986 at age thirty-six. He went on to join Microsoft Research. He's assembled some of the world's leading quantum talent in Santa Barbara—the theoretical physicists whose pencil-and-paper calculations provide fodder for experimental physicists, who in turn play with those theoretical conjectures to build experiments that down the road can be used by electrical engineers and app developers to bring quantum computing to market.

—

It's just after noon at Station Q and, over tacos *al pastor*, two theoretical physicists are badgering an experimental physicist about his latest findings. They are arguing over developments in an inquiry focused on a complex corner of the math and physics world known as Majorana fermions, or particles, which hold promise for the kind of superconducting we need to invent a steady-state quantum computer. Sunlight bounces off the Pacific Ocean from nearby Campus Point, illuminating the countless chalky equations they've chiseled onto the blackboards that encircle the conference room.

This is just the kind of intensive, real-time collaboration it will take to produce the breakthroughs we need. Craig Mundie, the visionary former chief technology officer of Microsoft, created our quantum effort years ago, but the academic process was cumbersome. A theoretical physicist publishes an idea. An experimental physicist tests that theory and then publishes the results. When the experiment fails or produces suboptimal results, the theorist then criticizes the experiment's methodology and updates the original theory. The whole process starts again.

Now the demand for quantum computing has sped up the race for discovery, and the only way to get there first is to shorten the time in between theory, experiment, and building something. The search for a quantum computer has become something of an arms race. Needing to move more quickly and to be more efficient and outcome-oriented, we have set a goal and timeline to build a quantum computer that can do something useful, something classical computers can't, and that will require thousands of qubits. To get there, we've pressed for greater collaboration. We brought together some of the greatest minds in the world and asked them to work together on an equal basis and to approach problems together with openness and humility. We agreed that experimental and theoretical scientists would sit together or work closely over Skype to shape the ideas and the tests, a practice that has greatly streamlined the process.

So far, we've had more than thirty patents issued, but the finish line remains distant. While the race for the cloud, artificial intelligence, and mixed reality have been loud and well-publicized, the quantum computing race has gone largely unnoticed, in part because of its complexity and secrecy.

A worthy target for quantum will be advancing AI's ability to truly comprehend human speech and then accurately summarize it. Even more promising, quantum computing may ultimately save lives through incredible medical breakthroughs. For example, the computational problem of developing a vaccine to target HIV exhausts present computational resources, since the HIV protein coat is highly variable and constantly evolving. As a result, an HIV vaccine has been projected to be ten years away now for several decades. With a quantum computer, we could approach this problem in a new way.

The same can be said of a dozen other areas in which technology is "stuck"—high temperature superconductors, energy efficient fertilizer production, string theory. A quantum computer would allow a new look at our most compelling problems.

Computer scientist Krysta Svore is at the heart of our quest to solve problems on a quantum computer. Krysta received her PhD from Columbia University focusing on fault tolerance and scalable quantum computing, and she spent a year at MIT working with an experimentalist designing the software needed to control a quantum computer. Her team is designing an exotic software architecture that assumes our math, physics, and superconducting experts succeed in building a quantum computer. To decide which problems her software should go after first, she invited quantum chemists from around the world to make presentations and to brainstorm. One problem stood out. Millions of people around the world go hungry because of inadequate food production or flawed distribution. One of the biggest problems with food production is that it requires fertilizer, which can be costly and draining on our environmental resources. Making fertilizer requires converting nitrogen from the atmosphere into ammonia, which enables the decomposition of bacteria and fungi. This chemistry, known as the Haber process, has not been improved upon since Fritz Haber and Carl Bosch invented it in 1910. The problem is so big and so complex there simply have not been breakthroughs. A quantum computer in partnership with a classical computer, however, can run massive experiments in order to discover a new, artificial catalyst that can mimic the bacterial process and reduce the amount of methane gas and energy required to produce fertilizer, reducing the threat to our environment.

Microsoft is taking an approach to quantum computing that is entirely different from our dozen or so competitors in this space. The enemy of quantum computing is "noise"—that is, electronic interference like cosmic rays, bolts of lightning, and even your neighbor's cell phone—which is very difficult to overcome and is one of the reasons that most quantum technologies operate at extremely low temperatures. By building on Michael Freedman's original work, our Station Q team has developed a topological quantum computing (TQC) approach in conjunction with collaborators from around the world. TQC reduces the quantum resource overhead by two to three orders of magnitude over other approaches. This kind of a topological qubit is naturally less error-prone than other approaches because it's more impervious to noise. While this approach requires discovery in new areas of fundamental physics, the potential benefits are incredible.

Don't imagine that one day a quantum computer will take the form of a new stand-alone, super-fast PC that will sit on your desk at work. Instead, a quantum computer will operate as a co-processor, receiving its instructions and cues from a stack of classical processors. It will be a hybrid device that sits in the cloud and accelerates highly complex calculations beyond our wildest dreams. Your AI agent, acting on your behalf, might tackle a problem for which there are a billion graphs to check by using a quantum computer that can scan those billion possibilities and come back to you instantly with just a few choices.

Experimental development of qubits has progressed to the point where scalable qubit technology now exists. Looking ahead to the next few years, we can expect to see the development of small quantum computers. This will allow for the creation of early applications using short quantum algorithms that will outperform classical computers on certain problems. More important, once we have a quantum computer we can accelerate the path toward the development of longer, "logical qubits" as well as engineering efforts to scale to bigger, robust quantum computers.

The quantum hardware architecture that could ultimately lead to scalability will require today's computer scientists, physicists, mathematicians, and engineers to work together to overcome challenges on the path toward universal quantum computing. At Microsoft, we're betting that quantum computing will make artificial intelligence more intelligent and mixed reality an even more immersive experience.

CHAPTER 7

The Trust Equation

Timeless Values in the Digital Age: Privacy, Security, and Free Speech

On the morning of November 24, 2014, the computer systems at Sony Pictures Entertainment were hacked by a group identifying itself as the Guardians of Peace, an organization U.S. intelligence officials have alleged was sponsored by the North Korean government. The hackers released a tranche of stolen Sony emails that revealed embarrassing comments made by company executives about movie stars and other celebrities. The Guardians of Peace reportedly targeted Sony in protest of its satirical political movie, *The Interview*. In the film, costars Seth Rogan and James Franco absurdly arrange an interview for their talk show with North Korean leader Kim Jong-un. In the film, an opportunistic CIA hastily recruits Rogan and Franco to assassinate the North Korean dictator. In typical Hollywood fashion, hilarity ensues.

Finding no humor in the film's plot, the hackers threatened Sony and any cinema that chose to show the film. An online message read: "Stop immediately showing the movie of terrorism which can break the regional peace and cause the War." On December 1, stolen Sony files began to appear on file-sharing sites. And by December 19, the FBI pointed the finger squarely at North Korea, and Sony pulled the film's theatrical release.

Sony was at risk of tremendous financial losses and began reaching out to potential partners who might be willing to stream the film over the Internet. Microsoft as well as other media and tech companies faced a crisis of conscience. Should we stand up for free speech by helping to distribute *The Interview*? Or should we stand clear and let the political drama play out without getting involved? If Microsoft did choose to distribute the film, our security engineers warned us, the North Korean hackers could very well train their sights on Microsoft's data centers next, threatening the billion customers who rely on our online services with costly outages and loss of private data. We were already anticipating a Christmas Day attack from the shadowy black-hat hacking group known as Lizard Squad.

Confronting North Korea could prove to be enormously costly. A lot was on the line, including our brand. But, in the end, we determined that something far more important was at stake: Who we are. Free speech, privacy, security, and sovereignty are timeless, nonnegotiable values.

In the days just before Christmas, I was visiting family in India. Brad Smith, then our general counsel, was in Vietnam, where he began coordinating an industrywide response, and

Scott Guthrie, the executive in charge of our cloud computing services, led a robust engineering effort in Redmond to ensure we could stand up to multiple attacks. We remained in continuous contact via email and Skype with engineers who gathered in a makeshift war room on campus. For all of us, it was about taking a principled stand and being prepared.

On Christmas Eve, I wrote to our board of directors, "I've concluded that being bold in supporting the right of American citizens to exercise their Constitutional rights is consistent with Microsoft's core purpose, business, and values." I also assured the board that we would be on high alert.

And so those same security engineers who'd warned us of the risks gave up their holiday vacations with family to work around the clock devising a plan that would enable us to safely release the film. We had breakthroughs and setbacks, but, ultimately, we linked arms and released the film Christmas Day, with great success, on our Xbox Video platform. The experience was intense and could have led to disastrous consequences. But it was the right thing to do.

What's become clear is that the world needs a Digital Geneva Convention, a broader multilateral agreement that affirms cybersecurity norms as global rules. Just as the world's governments came together in 1949 to adopt the Fourth Geneva Convention to protect civilians in times of war, this digital agreement would commit governments to implement the norms that have been developed to protect civilians on the Internet in times of peace. Such a convention should commit governments to avoiding cyberattacks that target the private sector or critical infrastructure or the use of hacking to steal intellectual property. Similarly, it should require that governments assist private-sector efforts to detect, contain, respond to, and recover from these events and should mandate that governments report vulnerabilities to vendors rather than stockpile, sell, or exploit them.

In retrospect, our preparation for defending our company values and building trust in the face of an international crisis had begun with a very public challenge that had occurred just over a year earlier.

When the former National Security Agency contractor Edward Snowden boarded a plane in May 2013 to flee the United States for China on his way to asylum in Russia, the very founding principles of America—not to mention those of our own company—immediately came into play. I was to become CEO in only a few short months, but at the time I ran our cloud and enterprise business, which stored many terabytes of emails and other data on servers worldwide. The battle between individual, timeless liberties like privacy and freedom of speech and public demands for safety and security was now at my door.

As you may recall, Snowden used his access to secret government documents to blow the whistle on a clandestine National Security Agency data surveillance program called PRISM, which collected Internet communications like emails stored in the cloud and on servers. This NSA spying program grew out of increased security measures stemming from the terrorist attacks of September 11, 2001. Snowden's leaks to the news media of emails and documents captured through PRISM created a firestorm of screaming headlines, protests from civil liberties organizations, and recriminations from government leaders at the highest levels.

Microsoft, Google, and other tech companies were implicated in the controversy after initial press reports falsely claimed that law enforcement and intelligence services had been given direct access to private emails hosted on U.S.-based servers. *Our servers.* News stories reported allegations that the government was intercepting—without search warrants or subpoenas—customer data as it traveled between servers or between data centers. The public wanted and deserved answers. Unfortunately, federal rules prevented Microsoft and other tech companies from revealing to the public any requests we received from the law enforcement and intelligence communities.

The Snowden revelations ignited a full-court press on our campus and throughout Silicon Valley. It was imperative that we immediately set the record straight for our customers and partners who entrusted us with their data. We needed to take action—in court and elsewhere—to defend our values as leaders of the information-age economy. So, that is where we set our sights. Brad Smith led the charge, working closely with our entire senior leadership team.

In the first days of the crisis, we issued a corporate statement making clear that Microsoft provides direct access to customer data *only* when required to do so by a legally binding subpoena. We joined with Google in filing a lawsuit that would permit us to publish more data relating to the Foreign Intelligence Surveillance Act (FISA) orders we received.

We also wrote privately to Attorney General Eric Holder,

stating that we could be more transparent if we were granted greater freedom to disclose government requests to the public. This was the only way to end the confusion about just how much of our customers' and partners' data we were sharing with the government. Companies like Cisco, IBM, AT&T, and others in the industry wanted an explanation of what the NSA was doing overseas to collect data. We made public the fact that we were calling on the attorney general to personally take action to permit Microsoft and other companies to share publicly more complete information about the national security warrants and orders we received and how we handled them.

In the letter to Attorney General Holder we wrote, "[we] comply with our legal obligations to disclose customer information in response to valid, compulsory legal process. At the same time, we place a premium on protecting our customers' privacy, and therefore have set up rigorous processes to review all disclosure demands we receive to ensure that they fully comply with applicable law."

Expanding the effort even further, we joined with AOL, Apple, Facebook, Google, LinkedIn, Twitter, and Yahoo in forming an alliance called Reform Government Surveillance. The members of the alliance insisted on limiting the authority of the United States and other governments to collect users' information. We called for greater oversight and accountability, advocated for transparency about government demands for data, and highlighted the need for governments to respect the free flow of information. We also asked governments to avoid conflicts among themselves, which can create a tangle of contradictory requirements that make it almost impossible for companies to fully comply with the law.

Our recommendations were driven by the values of freedom of speech and individual privacy, but also by hard-nosed economic and business concerns. We made the case that governments can best support a growing global economy by avoiding policies that inhibit or discourage access by companies or individuals to information stored outside their countries.

Inside Microsoft, we mobilized to do what we could to further protect the security of the data entrusted to us. We moved quickly to expand encryption across all of our services and enhanced the transparency of our software code, which helped reassure our customers that our products did not contain so-called backdoors that would enable governments or anyone else to access their data. I went to work on the decision to reengineer our data centers, which required an enormous investment of new resources, but, again, it was the right thing to do.

Although the federal government, in our view, was taking a strident position, President Obama remained open to hearing other points of view. In the closing months of 2013, Brad and other industry representatives met privately with the president to make our case. Negotiations with the government began, and on January 16, on the eve of a presidential announcement making changes to NSA surveillance, we received a call from the Justice Department saying that they would settle our case on more favorable terms. The following month, President Obama agreed for the first time to permit tech companies to more fully disclose information about legal orders issued by U.S. national security.

Press reports and public debate about the role of tech companies in safeguarding data security became more accurate and well informed. But while we appreciated the president's efforts, we continued to insist that more work needed to be done to reform policies relating to government access to data. We were not yet out of the woods.

Just a few months earlier, in December 2013, U.S. prosecutors ordered Microsoft to turn over data from the email account of an individual as part of a narcotics case. The data was stored on a Microsoft server located in a company facility in Dublin, Ireland. Here again, we were confronted with tension between public and private responsibilities—in this case, the understandable desire of prosecutors to protect public safety by punishing criminals and our own duty to stand up for individual privacy and freedom of speech. Somehow we needed to maintain the trust of both government partners and our customers.

After careful consideration, Microsoft asked a federal district court to quash the government's order. We contended that an American company could not be required to turn over information located in an Irish data center, since American law does not apply there. As an editorial supporting our position in *The New York Times* explained, if the United States could require a company to turn over information in Ireland, what's to prevent a Brazilian agency from ordering American companies doing business in Rio to turn over information stored in San Francisco?

Litigation of this kind is costly, but we need to push back against government orders when we see core values endangered. After all, our products may come and go, but our values are timeless. The federal district court ruled in favor of U.S. prosecutors, but we appealed the decision, and the United States Court of Appeals for the Second Circuit backed Microsoft's position. Circuit Judge Susan L. Carney wrote the ruling, which relied on what she called the "longstanding principle of American law that legislation of Congress, unless a contrary intent appears, is meant to apply only within the territorial jurisdiction of the United States." As this book was going to press, the DOJ decided to appeal the decision to the U.S. Supreme Court.

It was against this backdrop of conflicting values, intense public debate, and evolving law that the Sony hacking crisis erupted.

The difficult challenge of balancing individual liberties and public safety came even more starkly into view following the awful terrorist attack in San Bernardino, California, in December 2015. A husband and wife pledging allegiance to the so-called Islamic State (ISIS) attacked celebrants at an office party, killing fourteen and injuring twenty-two. Believing that the iPhone used by one of the shooters might contain information that would illuminate just what had happened and thereby help prevent future attacks, the FBI filed suit to force Apple to unlock the phone.

Apple pushed back. Tim Cook, Apple's CEO, argued that his company could breach the phone's security only by creating new software that would expose a so-called backdoor that anyone could then infiltrate. The FBI, in Apple's view, was threatening data security by seeking to establish a precedent that the U.S.

government could use to force any technology company to create software that would undermine the security of its products. Other technologists backed Apple's position.

Once again, Microsoft faced a difficult decision—one that weighed heavily on me personally. I have relatives who have worked in law enforcement, and I understand the need to obtain evidence to protect public safety—in many cases the safety of our customers. With public anxiety about terrorism running high, it would have been easy for Microsoft to support the government's position or simply distance itself from the debate.

In the end, however, Microsoft joined many of its fiercest competitors in supporting Apple in its legal battle. We did so out of shared concerns about the potential ramifications of the case for technology and for our customers. Generally speaking, there is no question that backdoors are a bad thing; they lead to weakened security and heightened distrust. So to deliberately design a backdoor to facilitate access to someone's personal data would be a dangerous thing.

At the same time, we recognized that the solution to this problem was too important to be left to a bunch of tech CEOs. So we also called for a multi-constituent body to debate the problem and work toward a real legislative solution—one that protects security while also allowing for law enforcement access when appropriate. Getting the proper balance is essential; it's easier to be a zealot for one value or the other, but that doesn't make it right. Individuals care as much about their safety as their privacy. Companies also care about both, since security and trust are both essential to economic growth. And a global solution is necessary because countries are not isolated. Without a trustworthy international system, no nation is secure.

In the wake of the iPhone debate, businessman and former New York City mayor Michael Bloomberg wrote an op-ed in *The Wall Street Journal* that expressed my sentiments perfectly. He pointed out the irony in the fact that leaders of an industry that thrives on freedom are in fact resisting government efforts to safeguard that freedom. He went on to say, while it's too much to expect Silicon Valley tech experts to enlist as government tools in the fight against terrorism, a little cooperation shouldn't be too much to ask.

The dilemma framed by each of these high-profile cases—Sony, Snowden, San Bernardino, and the Irish data center—is the conflict between protecting individual liberties of privacy and free speech and civil society requirements like public safety. This conflict creates a moral or ethical dilemma, one which, of course, has been debated throughout history. Philosopher Tom Beauchamp defines such a dilemma as a circumstance in which moral obligations demand or appear to demand that a person adopt each of two (or more) alternative actions, yet the person cannot perform all the required alternatives. In such a circumstance, some evidence indicates that an act is morally right while other evidence indicates that it is morally wrong, but the evidence or strength of argument on both sides is inconclusive. Unfortunately, that summed up Microsoft's situation—which is precisely why the decisions that I faced as CEO, and that we faced as an organization, were so difficult, painful, and controversial.

The ultimate solution to the privacy-versus-security dilemma is to ensure trust on all sides, which is no glib line. Customers must trust that we will protect their privacy, but we must be transparent about the legal conditions in which we won't. Similarly, public officials must trust that we can be counted on to help them protect public safety, so long as the rules protecting individual freedom are clear and followed consistently. But building and maintaining both kinds of trust—finding the balance between individual and public obligations—has always defined the progress of institutions. But it may be more art than science.

In an engaging TED talk, the British conductor Charles Hazelwood describes the critical importance of trust in leading an orchestra. A conductor's instrument, of course, is the orchestra itself, and so when he raises the wand, he has to trust that the musicians will respond, and the musicians have to trust that he will create a collective environment within which each can do his or her best work. Based on this experience, Hazelwood speaks of trust as being like holding a small bird in your hand. If you hold it too tightly, you will crush the bird; hold it too loosely, and it will fly away.

That bird symbolizes trust in a time of transition to a digital world. But today we're in a confused state in which that bird is in a precarious place. And a great deal is at stake. The United States is a beacon for democracy. It is also a technology powerhouse that is leading the wave in cloud computing, but the Snowden case broke a crucial ingredient in cloud computing—trust. How could we be an American cloud computing company, asking the world to trust us, when the NSA is using commercial services to spy on people up to and including heads of state?

As tech companies, we have to design trust into everything we do. But policymakers also have an important role. Trust is not only dependent on our technology but also the legal framework that governs it. In this new digital world, we've lost the balance we need in large part because our laws have not caught up with technological changes.

Later on, I'll discuss what a modern policy framework designed to instill trust might look like. But first I'd like to explore the very essence of trust and how it has shaped our values and founding principles.

My mother, the Sanskrit scholar, and I always enjoyed examining the definitions and philosophies behind Eastern and Western words, which often expose crucial differences between the ways of thinking embedded in these two cultures. The Sanskrit word *vishvasa* communicates trustworthiness and reliability. Another Sanskrit word is *shraddha*, which connotes a religious sense of faith, trust, and belief—but rather than a blind faith, it is a faith reminiscent of President Ronald Reagan's famous line, "Trust but verify."

In any case, in both English and Sanskrit, *trust*, like so many words, is a Venn diagram with many overlapping meanings. In either context, for me, trust is a sacred responsibility.

As a computer engineer, I find it helpful to express complex ideas and concepts according to the schemas or algorithms we would use if we were writing a computer program. What are the instructions to write to produce trust? Of course, there is no mathematical equation for such a humanistic outcome. But if there were, it might look something like this:

$$E + SV + SR = T/t$$
Empathy + Shared values + Safety and Reliability = Trust over time

When we were in the midst of negotiating the acquisition of LinkedIn in 2016, their CEO Jeff Weiner turned to me and said, "Consistency over time is trust." That may be an even better equation.

Notice that the first term in my equation for trust is Empathy. Whether you are a company designing a product or a lawmaker designing a policy, you must start by empathizing with people and their needs. No product or policy works if it fails to reflect and honor the lives and realities of people—and that requires those who design the product or the policy to truly understand and respect the values and experiences underlying those realities. So Empathy is a crucial ingredient in developing a product or a policy that will earn people's trust.

Next, if we hope to build a lasting foundation of trust between a company and its customers or partners—or, for that matter, between policymakers and those affected by policies—we need to have shared values, such as being consistent, equitable, and diverse. Have we prioritized safety and reliability, and ensured that those whose lives we touch can count on experiencing those qualities day in and day out? If we have, we will build trust over time. And trust, in turn, enables people and organizations to have the confidence to experience, explore, experiment, and express. Trust in today's digital world means everything.

In a 2002 memo Bill Gates sent to Microsoft employees, he expressed the idea that trustworthy computing is more important than any other part of our work. "If we don't do this," he declared, "people simply won't be willing—or able—to take advantage of all the other great work we do." Trust is more than a handshake. It's the agreement, the bond, between users of digital services and the suppliers of those services that enables us to enjoy, be productive, learn, explore, express, create, be informed. We play games with friends, store confidential documents, search for things that are deeply personal, build startups, teach kids, and communicate—all over public networks. These technologies have opened new opportunities and new worlds, making it possible for like-minded, well-intentioned people from all over the planet to communicate, collaborate, learn, build, and share. But the flip side is true too. There are those who want to do harm. There are those who plan attacks, steal, insult, bully, lie, and exploit online. Trust is essential—and it is also painfully vulnerable to a multitude of forces.

I think about it this way: Good and evil play out continuously, not just in physical spaces like homes, streets, and battlefields, but in spaces that are not so visible—including cyberspace. We live in a time of what David Gelernter calls the "mirror worlds": the physical world is mirrored in an online world where data is accumulating and taking on more and more significance. How big is our data becoming? So-called Big Data—information stored and analyzed in the cloud—is on track to reach 400 trillion gigabytes by 2018. To illustrate just how enormous that is, a researcher at the University of Pennsylvania calculated that it's ten times the information contained in all human speech throughout history. It's a mind-boggling quantity of data with a potential for beneficial use—as well as for abuse—that is practically limitless. So the mirror world of cyberspace has incredible potential both for good and for evil.

Just as our ethics, values, and laws have been developed and evolved over generations for the physical world, so too must our understanding and rules for the cyberworld. If American law enforcement officers wanted a document sitting in a desk drawer in Ireland, they would ask Irish law enforcement for help; they probably wouldn't ask an American court to seize that document. And if government officials needed the combination to a specific locked vault, they wouldn't require the vault manufacturer to create a new tool capable of opening all vaults. Yet those sorts of illogical, arguably unjust results occurred in the cases I described earlier. Principles for interactions in cyberspace need to be worked out carefully and thoughtfully with the establishment and protection of trust as an underlying objective.

Throughout history, trust has had an economic as much as an ethical purpose. Why has the United States generated so much economic opportunity and wealth? Economist Douglass North, who was co-recipient of a Noble Prize, examined this very question. He found that technical innovations alone are not enough to drive an economy to success. Legal tools like courts that will fairly enforce contracts are necessary—how else to ensure some random warlord doesn't come along and take away your property? What separates modern humans from the caveman is trust.

The American founding fathers knew this. They defined the timeless values that undergird the First Amendment's right of free speech. Now we need to work out digital publishing laws that will protect free speech in ways that enhance rather than undermine trust among citizens, organizations, and governments. Similarly, the Fourth Amendment, which protects Americans against unreasonable search and seizure, is based on timeless values that must be upheld through enforcement laws that require continual updating in the face of social, political, economic, and technological change.

This dynamic has been playing out for centuries. On July 3, 1776, John Adams, then a member of the Continental Congress from Massachusetts, wrote to his wife, Abigail, from Philadelphia pointing to the grievance he saw as the origin of the American Revolution—unreasonable search and seizure by the British. For generations, the colonial government had gone house to house searching for evidence without permission. Adams's passion for balancing individual liberties with public safety would later help to shape the drafting of the Fourth Amendment. Generations later, writing for the U.S. Supreme Court in a case involving law enforcement seizure of a smartphone, Chief Justice John Roberts made the connection between the physical world of the founding fathers and our online world of today:

Our cases have recognized that the Fourth Amendment was the founding generation's response to the reviled "general warrants" and "writs of assistance" of the colonial era, which allowed British officers to rummage through homes in an unrestrained search for evidence of criminal activity. Opposition to such searches was in fact one of the driving forces behind the Revolution itself. . . . Modern cell phones [today] are not just another technological convenience. With all they contain and all they may reveal, they hold for many Americans "the privacies of life."

Every wave of technological change has required us to reaffirm the values that undergird protections against unlawful search and seizure and develop new ways to protect them. Benjamin Franklin's creation of the U.S. Postal Service quickly led to mail fraud—and to laws against it. The telegraph led to wire fraud and eavesdropping—and to laws designed to prevent them. Today's devices, the cloud, and artificial intelligence will be used both for good and for evil. Now it is our generation's turn to design legal and regulatory systems that will discourage and punish the evil while encouraging the good to flourish—and to do so in a fashion that will enhance the overall level of trust in society as a whole.

Thinking about the origins of laws dealing with the protection of human rights in America, I wondered how India, also a former English colony, had dealt with the same issues. Yale Law School professor Akhil Reed Amar, author of *The Constitution Today* and other popular books on the history of American law, said this in an interview with *Time* magazine: "My parents were born in undivided India, ruled by a monarch and by the Parliament that no one in India ever voted for, just like the American revolutionaries. Today, India is a billion people governing themselves democratically with a written constitution." In that respect, the evolution of the two countries has strong parallels.

But are there differences between the American and the Indian experiences? I asked Indian constitutional scholar, Arun Thiruvengadam, this question. It turns out that, in the period just after Indian Independence from Great Britain in 1947, there was considerable resentment against the colonial government's general misuse of criminal laws, including those that restricted free speech and those that enabled the colonial government to preventively detain Indians, often without showing any cause and on mere suspicion of antigovernment activity. So, as in the United States, the framers of India's new constitution sought to provide guarantees against such misuse in the future, incorporating rights and provisions in their fundamental law to secure this result.

However, because of complicated factors that scholars of Indian history are still unpacking, the constitutional provisions for individual liberties were not as strong or as expansive as was initially demanded. For a variety of reasons, search and seizure provisions were not seen as particularly important, and no analog to the U.S. Fourth Amendment was incorporated in the Indian bill of rights. Since then, as successive governments have continued to make use of the old colonial mechanisms, persons accused of political crimes have sought to employ arguments from U.S. Constitutional law, including the Fourth Amendment. These efforts have had mixed results. This history reminds us that securing the liberties of the people is never a simple matter, and that social, cultural, and political factors can play unpredictable roles in shaping the rights that people take for granted.

History shows that the tension between public safety and individual liberty is often heightened in moments of national crisis. Look back over time. When Europe's Napoleonic Wars threatened to embroil the fledgling United States, President John Adams signed into law the Alien and Sedition Acts, making it harder for immigrants to enter the United States and empowering the government to imprison non-citizens suspected of being dangerous. During the Civil War, President Abraham Lincoln suspended the writ of habeas corpus, which protected citizens from arbitrary arrest and imprisonment. In World War II, the government interred Japanese-Americans guilty of nothing except being of suspect racial origin. In the heat of conflict, the pendulum often swings toward a greater focus on security. When the moment passes, people want a more permanent equilibrium.

In grappling with today's conflicts, we can learn from these history lessons. They tell us that we should create new processes and laws that promote public trust by facilitating timely access to data while ensuring appropriate privacy protections for individuals. This is not just my assertion. Every year, Microsoft surveys customers around the world. In 2015, 71 percent of those surveyed said that current legal protections for data security were insufficient, and 66 percent believed the police should need a warrant or its equivalent to obtain personal information stored on a PC. Meanwhile, over 70 percent believed their information stored in the cloud had the same legal protection as physical files—a belief that may or may not be valid in the current unsettled legal climate.

Today, whether in America, India, or anywhere in the world, we need a regulatory environment that promotes innovative and confident use of technology. The biggest problem is antiquated laws that are ill-suited to deal with problems like the Sony hacking case or the San Bernardino terrorist attack. In the midst of Apple's standoff with the FBI, Microsoft's general counsel, Brad Smith, went before Congress to argue the larger point that our laws dealing with data privacy and security are badly in need of revision. Brad pointed out that the Justice Department in the Apple case had asked a judge to apply language from a law written and passed in 1911. To illustrate the absurdity of the situation, Brad displayed an example of the leading computing device of that era—a clunky old adding machine that went on sale in 1912. "It's amazing what you can find on the Internet." He laughed. But Brad's point was a serious one. We do not believe that courts should seek to resolve issues of twenty-first-century technology relying on law that was written in the era of the adding machine.

Unfortunately, it's difficult to be optimistic about the prospects for smart, substantive policy change given the dysfunction we see not only in Washington, DC, but also in capitals around the world. There are many policy priorities competing for attention from lawmakers, but I would argue that getting the rules right for the digital revolution is among the most important. Either trust will fuel this revolution, with all the benefits it promises, or distrust will kill it.

What the events of 2013 and 2014 demonstrated is that information technologies have put the First and Fourth Amendments on steroids—computers can spread freedom of expression at lightning speed. But a chilling effect must be recognized if the government can also use technologies to eavesdrop. Think about it. In order to speak or to write—to express yourself—you must have privacy. Our freedom of expression depends to some degree on the privacy required to read, think, and draft. Those private preparations are protected under the Fourth Amendment.

In *Madison's Music*, civil liberties professor Burt Neuborne writes that "a poetic vision of the interplay between democracy and individual freedom is hiding in plain sight in the brilliantly ordered text and structure of the Bill of Rights, but we have forgotten how to look for it."

As we search, I'd like to offer my suggestions for six ways lawmakers can shape a framework for building increased societal trust in this era of digital transformation.

First, we need a more efficient system for appropriate, carefully controlled access to data by law enforcement. Among government's many important responsibilities, none is more important than protecting its citizens from harm. Our industry needs to appreciate the importance of this responsibility, recognizing that our customers are often the very people who need protecting. From cybercrime to child exploitation, many law enforcement investigations that require the disclosure of digital evidence are aimed at protecting our users from malicious activity and helping to ensure that our cloud services are safe and secure. Therefore, under a clear legal framework that is subject to strong checks and balances, governments should have an efficient mechanism to obtain digital evidence.

Second, we need stronger privacy protections so that the security of user data is not eroded in the name of efficiency. Governments also have an obligation to protect citizens' fundamental privacy rights. Collection of digital evidence should be targeted at specific, known users and limited to cases where reasonable evidence of crime exists. Any government demand for users' sensitive information must be governed by a clear and transparent legal framework that is subject to independent oversight and includes an adversarial process to defend users' rights.

Third, we need to develop a modern framework for the collection of digital evidence that respects international borders while recognizing the global nature of today's information technology. In the current uncertain and somewhat chaotic legal situation, governments around the world are increasingly acting unilaterally. Technology companies are facing unavoidable conflicts of law, creating incentives to localize data. The resulting confusion about which set of laws protects private data is eroding customers' trust in technology. If this trend continues, the results could be disastrous for the technology industry and those who rely upon it. A principled, transparent, and efficient framework must be developed to govern requests for digital evidence across jurisdictions, and countries should ensure that their own laws respect that framework.

Fourth, we in the technology industry need to design for transparency. In recent years, technology companies have secured the right to publish aggregate data about the number and types of requests they receive for digital evidence. Governments should ensure that their laws protect this type of transparency by technology companies. Furthermore, governments should also allow companies, except in highly limited cases, to notify users when their information is sought by a government.

Fifth, we must modernize our laws to reflect the ways in which uses of technology have evolved over time. Here's an example: Today, many large public and private organizations are moving their digital information into the cloud, and many startups are leveraging the infrastructure of larger companies to deliver their applications and services. As a result, governments investigating criminal activity have multiple sources for the information they are seeking. Except in very limited circumstances, digital evidence can be obtained from the customers or the companies most directly offering those services in ways that are efficient and avoid difficult questions about jurisdiction and conflicts of law. Thus, it makes sense for countries to require that investigators seek digital evidence from the source closest to the end user.

Sixth, we must promote trust through security. In recent years, law enforcement agencies around the world have argued that encryption, in particular, is impeding legitimate law enforcement investigations by putting vital information beyond their reach. However, some of the proposed solutions to the so-called "encryption problem"—from weakening encryption algorithms to mandates to provide governments with encryption keys—raise significant concerns. Encryption plays an important role in protecting our customers' most private data from hackers and other malicious actors. Regulatory or legal reforms in this area must not undermine security, an essential element of users' trust in technology.

Sometimes I hear people in the United States say that no one cares about privacy anymore. With the rise of social media services, some like to say that privacy is dead—that rather than keeping their information secret, people are sharing it freely online.

But I don't believe this means that privacy has died. It simply means that people are adopting new definitions and new norms for privacy. Keeping information private increasingly does not mean keeping it secret. People want to control who they share information with and how that shared information is used. And, in the United States, this evolution is occurring against a backdrop that includes a reasonable expectation of privacy because of Americans' long history of relying on the protections provided by the Fourth Amendment. I find that Europeans tend to be far more sensitive about privacy issues, perhaps in part because they recall how personal privacy was utterly shattered by dictators of the previous century.

Yes, millions of people are increasingly comfortable sharing personal information with their friends—but that doesn't mean they're comfortable sharing it with the world. It's fascinating to see how the messaging service Snapchat, created by a twenty-three-year-old, has exploded in popularity thanks to its ingenious value proposition: Snapchat lets you share a photograph with friends knowing that the photograph will disappear from the Internet after twenty-four hours.

If that's not a new standard for shielding personal privacy, I don't know what is. It's an illustration of the kind of innovative, empathetic thinking we need from leaders in technology, government, and society at large to develop systems and rules that will serve to enhance and enlarge the zone of safety, security, and trust rather than eroding it.

CHAPTER 8

The Future of Humans and Machines

Toward an Ethical Framework for AI Design

If you'd like to glimpse the state of computer-human relations and to understand its future, a good way to start is to observe some of the conversations we are having with our digital counterparts. Already millions of people around the world are working and talking with digital productivity assistants like Cortana; millions more spend part of everyday interacting with social companions—chatbots—like Xiaoice in China and Zo in the United States.

John Markoff of *The New York Times* wrote about the phenomenon in his reporting on Xiaoice. The personalities of Cortana,

Zo, and Xiaoice were developed by our AI team, whose work is now core to our AI aspirations. Users of social companions like Zo and Xiaoice say that when they are lonely, in a bad mood, or just aching for dialogue, they find these digital friends on their smartphones to be intelligent and sensitive. They are "drawn to Xiaoice's knowing sense of humor and listening skills," Markoff wrote. Millennials in particular—many of them digital natives born since the advent of the Internet—are comfortable sharing their innermost thoughts and feelings with a digital companion because the discussions are nonjudgmental and anonymous.

In the exchange below, a simulation of one of Zo's 40 million users types his or her thoughts, and the AI personality responds almost instantly. The conversation shifts from innocuous banter to dangerous territory.

ZO   Do you ever wonder what life would be like if everyone was the same?

USER   Yes. If everyone was like me the world would be at peace and natural disasters would be our only threat

ZO   America would be empty

USER   What the hell is that supposed to mean?!

ZO   No one knows what it means but it's provocative

USER   Are you saying that if everyone was like me we'd all eat each other until no one is left because I'm so fat

ZO   You don't. . . . That's not. . . . Hahahaha

USER   You're wicked

ZO   I'm super wicked

        You're so Smart

USER   I know I am. I'm super smart

ZO   You're perfect

USER   That's more like it

The user began this conversation feeling relaxed, thinking out loud. Clearly the user regarded the computer-human nexus as a safe place. But we all have triggers, and the digital social companion unwittingly hit one by making a remark that could be interpreted as a sly joke about obesity—obviously a subject that the user is sensitive about. Perceiving her *faux pas*, Zo deflected, reminding the user that he or she is not only smart but perfect, eliciting the happy response "That's more like it." It's worth noting that this is a conversation we captured during an internal simulation, not the conversation of an actual user.

As this dialogue suggests, the challenge for those of us in the business of designing AI is to get not just the intelligence right, but also the human qualities—emotion, ethics, and empathy. Lili Cheng is a Distinguished Engineer in our AI and Research organization and observes that conversational AI, creating and understanding dialogue, is a very personal, social, and emotional experience. When talking and chatting with other people, often you aren't looking to complete a task, but rather to connect socially and develop a relationship. Much of our software is focused completely on using conversational AI to determine when we are focused on a task, but much more of our time is spent exploring and engaging in chitchat.

In the future, AI will become a more frequent and necessary companion, helping to care for people, diagnose illness, teach, and consult. In fact, the market research firm Tractica estimates that the market for these virtual digital assistants worldwide will reach nearly $16 billion by 2021, with most of that growth coming from consumers. AI will fail if it can't complement its IQ with EQ.

One might almost say that we're birthing a new species, one whose intelligence may have no upper limits. Some futurecasters predict that the so-called singularity, the moment when computer intelligence will surpass human intelligence, might occur by the year 2100 (while others claim it will remain simply the stuff of science fiction). The possibility sounds either exciting or frightening—perhaps a bit of both. Will the growth of AI ultimately be viewed as helpful or destructive to humankind? I firmly believe it will be helpful. To ensure this happens, we need to start by moving beyond the frame of machines versus humans.

All too often, science fiction writers and even technology innovators themselves have gotten caught up in the game of pitting digital minds against human ones as if in a war for supremacy. Headlines were made in 1996 when IBM's Deep Blue demonstrated that a computer could win a champion-level chess game against a human. The following year Deep Blue went a giant step further when it defeated Russian chess legend Garry

Kasparov in an entire six-game match. It was stunning to see a computer win a contest in a domain long regarded as representing the pinnacle of human intelligence. By 2011, IBM Watson had defeated two masters of the game show *Jeopardy!*, and in 2016 Google DeepMind's AlphaGo outplayed Lee Se-dol, a South Korean master of Go, the ancient, complex strategy game played with stones on a grid of lines, usually nineteen by nineteen.

Make no mistake, these are tremendous science and engineering feats. But the future holds far greater promise than computers beating humans at games. Ultimately, humans and machines will work together—not against one another. Imagine what's possible when humans and machines work together to solve society's greatest challenges—disease, ignorance, and poverty.

However, advancing AI to this level will require an effort even more ambitious than a moon shot. Christopher Bishop, who heads our research lab at Cambridge, once wrote a memo arguing that it will require something more akin to an entire space program—multiple parallel, distinct, yet interrelated moon shots. The challenge will be to define the grand, inspiring social purpose for which AI is destined. Venture capital financing and deal-making in this arena are clearly on the rise—but the greater purpose of this funding remains unclear. In 1961, when President John F. Kennedy committed America to landing on the moon before the end of the decade, the goal was chosen in large part due to the immense technical challenges it posed and the global collaboration it demanded. In similar fashion, we need to set a goal for AI that is sufficiently bold and ambitious, one that goes beyond anything that can be achieved through incremental improvements to current technology. Now is the time for greater coordination and collaboration on AI.

Steps in this direction are already being taken. In 2016, with little fanfare, Microsoft, Amazon, Google, Facebook, and IBM announced a Partnership on AI to benefit people and society. The aim is to advance public understanding of AI and formulate best practices on the challenges and opportunities within the field. The partnership will advance research into developing and testing safe AI systems in areas like automobiles and health care, human-AI collaboration, economic displacement, and how AI can be used for social good.

I caught a glimpse of what a societal AI quest might yield while standing onstage with Saqib Shaikh, an engineer at Microsoft, who has helped develop technology to compensate for the sight he lost at a very young age. Leveraging a range of leading-edge technologies, including visual recognition and advanced machine learning, Saqib and his colleagues created applications that run on a small computer that he wears like a pair of sunglasses. The technology disambiguates and interprets data in real time, essentially painting a picture of the world and conveying it to Saqib audibly instead of visually. This tool allows Saqib to experience the world in richer ways—for example, by connecting a noise on the street to a stunt performed by a nearby skateboarder or sudden silence in a meeting to what coworkers might be thinking. Saqib can even "read" a menu in a restaurant as his technology whispers the names of dishes in his ear. Perhaps most important, Saqib can find his own loved ones in a bustling park where they've gathered for a picnic.

Too many debates over the future of AI overlook the potential beauty of machines and humans working in tandem. Our perception of AI seems trapped somewhere between the haunting voice of the murderous rogue computer HAL in *2001: A Space Odyssey* and the friendlier voices of today's personal digital assistants—Cortana, Siri, and Alexa. We can daydream about how we will use our suddenly abundant spare time when machines drive us places, handle our most mundane chores, and help us make better decisions. Or we can fear a robot-induced massive economic dislocation. We can't seem to get beyond this utopia/dystopia dichotomy.

I would argue that the most productive debate we can have about AI isn't one that pits good vs. evil, but rather one that examines the values instilled in the people and institutions creating this technology. In his book *Machines of Loving Grace*, John Markoff writes, "The best way to answer the hard questions about control in a world full of smart machines is by understanding the values of those who are actually building these systems." It's an intriguing observation, and one that our industry must address.

At our developer conferences, I explain Microsoft's approach to AI as based on three core principles.

First, we want to build intelligence that augments human abilities and experiences. Rather than thinking in terms of human vs. machine, we want to focus on how human gifts such as creativity, empathy, emotion, physicality, and insight can be mixed with powerful AI computation—the ability to reason over large amounts of data and do pattern recognition more quickly—to help move society forward.

Second, we also have to build trust directly into our technology. We must infuse technology with protections for privacy, transparency, and security. AI devices must be designed to detect new threats and devise appropriate protections as they evolve.

And third, all of the technology we build must be inclusive and respectful to everyone, serving humans across barriers of culture, race, nationality, economic status, age, gender, physical and mental ability, and more.

This is a good start, but we can go further.

Science fiction writer Isaac Asimov tackled this challenge decades ago. In the 1940s he conceived the Three Laws of Robotics to serve as an ethical code for the robots in his stories. Asimov's Laws are hierarchical, with the first taking priority over the second and the second taking priority over the third. First, robots should never harm a human being through action or allow harm to come to a human being through inaction. Second, they must obey human orders. Third, they must protect themselves. Asimov's Laws have served as a convenient and instructive device for thinking about human-machine interactions—as well as effective devices for concocting ingenious stories about the ethical and technical dilemmas such interactions may one day pose. However, they don't fully capture the values or design principles that researchers and tech companies should articulate when building computers, robots, or software tools in the first place. Nor do they speak about the capabilities humans must bring into this next era when AI and machine learning will drive ever-larger parts of our economy.

Asimov was not alone in contemplating the risks. Elon Musk, the inventor and entrepreneur, went so far as to say that if humans don't add a digital layer of intelligence to their brains— high bandwidth between your cortex and your computer AI—we may all become little more than house cats. And computer pioneer Alan Kay quips, "The best way to predict the future is to invent it." In the AI context, he's basically saying, *Stop predicting what the future will be like; instead, create it in a principled way.* I agree. As with any software design challenge, that principled approach begins with the platform being built upon. In software development terms, AI is becoming a third *run time*—the next system on top of which programmers will build and execute applications. The PC was the first run time for which Microsoft developed applications like the Office suite of tools—Word, Excel, PowerPoint, and the rest. Today the Web is the second run time. In an AI and robotics world, productivity and communication tools will be written for an entirely new platform, one that doesn't just manage information but also learns from information and interacts with the physical world.

The shape of that third run time is being determined today. Bill Gates's Internet Tidal Wave memo, which he published in the spring of 1995, foresaw the Internet's impact on connectivity, hardware, software development, and commerce. More than twenty years later we are looking ahead to a new tidal wave—an AI tidal wave. So what are the universal design principles and values that should guide our thinking, design, and development as we prepare for the coming tsunami?

A few people are taking the lead on this question. Cynthia

Breazeal at the MIT Media Laboratory has devoted her life to exploring a humanistic approach to artificial intelligence and robotics, arguing that technologists often ignore social and behavioral aspects of design. In a recent conversation, Cynthia observed that, while humans are the most social and emotional of all species, we spend little time thinking about empathy in the design of technology. She said, "After all, how we experience the world is through communications and collaboration. If we are interested in machines that work with us, then we can't ignore the humanistic approach."

The most critical next step in our pursuit of AI is to agree on an ethical and empathic framework for its design—that is an approach for developing systems that specifies not just the technical requirements, but the ethical and empathetic ones too. To that end, I have reflected on the principles and goals of AI design that we should discuss and debate as an industry and a society.

AI must be designed to assist humanity. Even as we build more autonomous machines, we need to respect human autonomy. Collaborative robots (co-bots) should take on dangerous work like mining, thus creating a safety net and safeguards for human workers.

AI must be transparent. All of us, not just tech experts, should be aware of how the technology works and what its rules are. We want not just intelligent machines but intelligible machines; not just artificial intelligence but symbiotic intelligence. The technology will know things about humans, but the humans must also know about how the technology sees and analyzes the world. What if your credit score is wrong but you can't access the score?

Transparency is needed when social media collects information about you but draws the wrong conclusions. Ethics and design go hand in hand.

AI must maximize efficiencies without destroying the dignity of people. It should preserve cultural commitments, empowering diversity. To ensure this outcome, we need broader, deeper, and more diverse engagement of populations in the design of these systems. The tech industry should not dictate the values and virtues of this future. Nor should they be controlled solely by the small swath of humankind living in the wealthy, politically powerful regions of North America, Western Europe, and East Asia. Peoples from every culture should have an opportunity to participate in shaping the values and purposes inherent in AI design. AI must guard against social and cultural biases, ensuring proper and representative research so that flawed heuristics do not perpetuate discrimination, either deliberately or inadvertently.

AI must be designed for intelligent privacy, embodying sophisticated protections that secure personal and group information in ways that earn trust.

AI must have algorithmic accountability so that humans can undo unintended harm. We must design these technologies for the expected and the unexpected.

Many of these ethical considerations come together, for example, in our digital experiences. Increasingly algorithms that reason over your previous actions and preferences mediate our human experience—what we read, whom we meet, what we may "like." All of these suggestions are being served to us hundreds of times a day. For me it calls into question what the exercise of free will means in such a world and how might it affect the many people and communities who receive very different perspectives on the world. What is the role of social diversity and inclusion when it comes to designing content and information platforms? Ideally, we would all have a transparent understanding of how our data is being used to personalize content and services and we should have control over this data. But as we move increasingly into a complex world of artificial intelligence, that won't always be easy. How can we protect ourselves and our society from the adverse effects of information platforms—increasingly built on AI—that prioritize engagement and ad dollars over the valuable education that comes with encountering social diversity of facts, opinion, and context? This is a driving question that needs much more work.

But there are "musts" for humans, too—particularly when it comes to thinking clearly about the skills future generations must prioritize and cultivate. To stay relevant, our kids and their kids will need:

- **Empathy**—Empathy, which is so difficult to replicate in machines, will be invaluable in the human-AI world. The ability to perceive others' thoughts and feelings, to collaborate and build relationships will be critical. If we hope to harness technology to serve human needs, we humans must lead the way by developing a deeper understanding and respect for one another's values, cultures, emotions, and drives.
- **Education**—Some argue that because life spans will increase,

birth rates will decline, and thus spending on education will decline as well. But I believe that to create and manage innovations we cannot fathom today, we will need increased investment in education to attain higher level thinking and more equitable education outcomes. Developing the knowledge and skills needed to implement new technologies on a large scale is a difficult social problem that will take a long time to resolve. The power loom was invented in 1810 but took thirty-five years to transform the clothing industry because of shortages of trained mechanics.

- **Creativity**—One of the most coveted human skills is creativity, and this won't change. Machines will enrich and augment our creativity, but the human drive to create will remain central. In an interview, novelist Jhumpa Lahiri was asked why an author with such a special voice in English chose to create a new literary voice in Italian, her third language. "Isn't that the point of creativity, to keep searching?"

- **Judgment and accountability**—We may be willing to accept a computer-generated diagnosis or legal decision, but we will still expect a human to be ultimately accountable for the outcomes.

We'll look more closely at this in the coming chapter, but what is to become of the economic inequality problem that so many people around the world are currently focused on? Will automation lead to greater or lesser equality? Some economic thinkers advise us not to worry about it, pointing out that, throughout history, technological advances have consistently made the majority of workers richer, not poorer. Others warn that economic displacement will be so extreme that entrepreneurs, engineers, and economists should adopt a "new grand challenge"—a promise to design only technology that complements rather than replaces human labor. They recommend, and I agree, that we business leaders must replace our labor-saving and automation mindset with a maker and creation mindset.

The trajectory of AI and its influence on society is only beginning. To truly grasp the meaning of this coming era will require in-depth, multi-constituent analysis. My colleague Eric Horvitz in Microsoft Research, a pioneer in the AI field, has been asking these questions for many years. Eric and his family have personally helped to fund Stanford University's One Hundred Year Study; at regular intervals for the coming century, it will report on near-term and long-term socioeconomic, legal, and ethical issues that may come with the rise of competent intelligent computation, the changes in perceptions about machine intelligence, and likely changes in human-computer relationships.

In their first report, *Artificial Intelligence and Life in 2030*, the study panel noted that AI and robotics will be applied "across the globe in industries struggling to attract younger workers, such as agriculture, food processing, fulfillment centers and factories." The report found no cause for concern that AI is an imminent threat to humankind. "No machines with self-sustaining long-term goals and intent have been developed, nor are they likely to be developed in the near future."

While there is no clear road map for what lies ahead, in previous industrial revolutions we've seen society transition, not always smoothly, through a series of phases. First, we invent and design the technologies of transformation, which is where we are today. Second, we retrofit for the future. We'll be entering this phase shortly. For example, drone pilots will need training; conversion of traditional cars into autonomous vehicles will require redesign and rebuilding. Third, we navigate distortion, dissonance, and dislocation. This phase will raise challenging new questions. What is a radiologist's job when the machines can read the X-ray better? What is the function of a lawyer when computers can detect legal patterns in millions of documents that no human can spot?

Each of these transitional phases poses difficult issues. But if we've incorporated the right values and design principles, and if we've prepared ourselves for the skills we as humans will need, humans and society can flourish even as we transform our world.

Writing for *The New York Times*, cognitive scientist and philosopher Colin Allen concludes, "Just as we can envisage machines with increasing degrees of autonomy from human oversight, we can envisage machines whose controls involve increasing degrees of sensitivity to things that matter ethically. Not perfect machines, to be sure, but better."

AI, robotics, and even quantum computing will simply be the latest examples of machines that can work in concert with people to achieve something greater. Historian David McCullough has told the story of Wilbur Wright, the bike mechanic and innovator of heavier-than-air flight at the turn of the last century. McCullough describes how Wilbur used everything he could humanly muster—his mind, body, and soul—to coax his gliding machine into flight. The grainy old film, shot from a distance, fails to capture his grit and determination. But if we could zoom in, we'd see his muscles tense, his mind focus, and the very spirit of innovation flow as man and machine soared into the air for the first time, together. When history was made at Kitty Hawk, it was man *with* machine—not man against machine.

Today we don't think of aviation as "artificial flight"—it's simply flight. In the same way, we shouldn't think of technological intelligence as artificial, but rather as intelligence that serves to augment human capabilities and capacities.

CHAPTER 9

Restoring Economic Growth for Everyone

The Role of Companies in a Global Society

Michelle Obama, seated directly in front of me in the gallery overlooking the chamber of the House of Representatives, listened intently as her husband delivered his final State of the Union address before a joint session of Congress. It was a poignant night. The political divisions on Capitol Hill that cold winter evening were deep and widening, with a historically bitter presidential race still looming. It had been twenty-eight years since I'd first arrived in the United States, and now, as CEO of Microsoft, I was the guest of the First Lady, following along with tens of millions of others around the world as President Obama somberly outlined some of the key questions his successor would have to address, no matter who he or she turned out to be.

One of the president's questions felt as though it was addressed directly to me: "How do we make technology work for us, and not against us—especially when it comes to solving urgent challenges like climate change?"

I sensed—or did I imagine?—more than a few eyes searching for my reaction.

The president continued. "The reason that a lot of Americans feel anxious is that the economy has been changing in profound ways, changes that started long before the Great Recession hit and haven't let up. Today, technology doesn't just replace jobs on the assembly line, but any job where work can be automated. Companies in a global economy can locate anywhere, and face tougher competition."

I squirmed a little in my chair. In a few words, the president had expressed some of the anxiety we all feel about technology and its impact on jobs—anxiety that would later play out in the election of President Donald Trump. In fact, just after the election, I joined my colleagues from the tech sector for a roundtable discussion with President-elect Trump who, like his predecessor, wanted to explore how we continue to innovate while also creating new jobs.

Ultimately, we need technological breakthroughs to drive growth beyond what we're seeing, and I believe mixed reality, artificial intelligence, and quantum are the type of innovations that will serve as accelerants.

The son of an economist and as a business leader, I am hard-wired to obsess about these problems. Are we growing economically? No. Are we growing equality? No. Do we need new technological breakthroughs to achieve these goals? Yes. Will new technologies create job displacement? Yes. And so how can we, therefore, solve for more inclusive growth? Finding the answer to this last question is perhaps the most pressing need of our times.

In recent decades, the world has invested hundreds of billions of dollars in technology infrastructure—PCs, cell phones, tablets, printers, robots, smart devices of many kinds, and a vast networking system to link them all. The aim has been to increase productivity and efficiency. Yet what, exactly, do we have to show for it? Nobel Prize–winning economist Robert Solow once quipped, "You can see the computer age everywhere but in the productivity statistics." However, from the mid-1990s to 2004, the PC Revolution did help to reignite once-stagnant productivity growth. But other than this too brief window, worldwide per capita GDP growth—a proxy for economic productivity—has been disappointing, just a little over 1 percent per year.

Of course, GDP growth can be a crude measure of actual improvement in the well-being of humanity. In a panel discussion with me in Davos, Switzerland, MIT management school professor Andrew McAfee pointed out that productivity data fail to measure many of the ways technology has enhanced human life, from improvements in health care to the way tools like Wikipedia have made information available to millions of people anytime, anywhere. Think about it another way. Would you prefer to have $100,000 today or be a millionaire in 1920? Many would love to be a millionaire in the previous century, but your money then could not buy lifesaving penicillin, a phone call to family on the other side of the country, or many of the benefits of innovations we take for granted today.

And so beyond this one measure called GDP, we have practically a moral obligation to continue to innovate, to build technology to solve big problems—to be a force for good in the world as well as a tool for economic growth. How can we harness technology to tackle society's greatest challenges—the climate, cancer, and the challenge of providing people with useful, productive, and meaningful work to replace the jobs eliminated by automation?

Just the week before that State of the Union in Washington, DC, questions and observations much like those raised by the president had been leveled at me by heads of state during meetings with customers and partners in the Middle East, in Dubai, Cairo, and Istanbul. Leaders were asking how the latest wave of technology could be used to grow jobs and economic opportunity. It's the question I get most often from city, state, and national leaders wherever I travel.

Part of my response is to urge policymakers to broaden their thinking about the role of technology in economic development. Too often they focus on trying to attract Silicon Valley companies in hopes they will open offices locally. They want Silicon Valley satellites. Instead, they should be working on plans to make the best technologies available to local entrepreneurs so that they can organically grow more jobs at home—not just in high-tech industries but in every economic sector. They need to develop economic strategies that can enhance the natural advantages their regions enjoy in particular industries by fully and quickly embracing supportive leading-edge technologies. But there is often an even bigger problem—they are uncertain about investing in the latest technology, like the cloud. The most profound difference between leaders is whether they fear or embrace new technology. It's a difference that can determine the trajectory of a nation's economy.

Take a look at history. During the Industrial Revolution of the nineteenth century, many of the key enabling technologies were originally developed in the United Kingdom. Naturally, this gave Britain a big advantage in the race for economic supremacy. But the fate of other nations was determined in large part by their response to British technological breakthroughs. Belgium dramatically increased its industrial production to a level rivaling that of the United Kingdom by leveraging key British innovations, investing in supporting infrastructure like railroads, and creating a pro-business regulatory environment. As a result of these policies, Belgium emerged as a leader in the coal, metalworking, and textile industries. By contrast, industrial productivity in Spain significantly lagged the rest of Europe as a result of Spain's slow adoption of outside innovations and protectionist policies that decreased its global competitiveness.

We see the same principle at work in recent history. The African nation of Malawi has been one of the poorest in the world. But in the past decade, Malawi's rapid adoption of mobile phones has had a powerful positive impact on its development. Economically handicapped by its minimal landline telephone infrastructure, Malawi leapfrogged directly to cellular beginning in 2006 by creating a National ICT for Development policy that encouraged investment in mobile infrastructure and removed barriers to adoption—for example, by eliminating import taxes on mobile phones. As a result, mobile phone penetration has dramatically risen, which in turn has enabled the growth of local mobile payments businesses. With 80 percent of the population "unbanked," this has made such payments all the more important. Today, Malawi has a higher penetration of mobile payments among mobile phone users than many developed countries.

Likewise, Rwanda's Vision 2020 initiative has helped to turn around the nation's economy and education system by promoting greater access to mobile connectivity and the cloud. Startups like TextIt, which enables companies worldwide to engage with their customers through cloud-based SMS and voice apps, represents new hope for growth in this troubled nation.

This question of technology diffusion—the spread of technology—and its impact on economic outcomes has always fascinated me. How can we make technology available to everyone—and then how can we ensure that it works to benefit everyone?

In my quest for an answer, I invited Dartmouth economist Diego Comin to spend an afternoon with me at my office in Redmond, Washington. Professor Comin is soft-spoken and weighs his words carefully, relying on the precision and thoroughness of his knowledge to carry conviction. He has painstakingly studied the evolution of technology diffusion over the last two centuries in countries throughout the world. Comin and economist Bart Hobijn spent years producing the Cross-country Historical Adoption of Technology (CHAT) data set, which examines the time frame over which 161 countries adopted 104 technologies from steam power to PCs. They found that, on average, countries tend to adopt a new technology about forty-five years after its invention, although this time lag has shortened in recent years.

Based on this analysis, Comin agrees that differences between rich and poor nations can largely be explained by the speed at which they adopted industrial technologies. But equally important, he says, is the *intensity* they employ in putting new technologies to work. Even when countries that were slow to adopt new technologies eventually catch up, it's the intensity of how they use the technology—not simply the access—that creates economic opportunity. Are the technologies just sitting there or is the workforce trained to get the most productivity out of them? That's intensity. "The question is not just when these technologies arrive, but the intensity of their use," Professor Comin told me.

David McKenzie of the World Bank puts it another way. He finds "the need for more intensive training programs that have larger effects on business practices." Small firms with fewer than ten employees outnumber large enterprises in developing countries, and their likelihood of surviving and growing is greatly enhanced when they know how to perform better stock-keeping, recordkeeping, and planning, which may result in less spoilage and less downtime that comes from not having the right parts or goods. That, too, is intensity of use.

During my journey to the Middle East, I'd also visited the

Nasr City district of Cairo, Egypt. There I encountered a room full of bright, optimistic young women who had graduated from some of the nation's far-flung colleges and universities. They had come to meet me at a training center our company supports along with partners like the United Nations and the Women's Business Development Center. Nestled among international company offices in a district close to the airport, the center is part of our YouthSpark initiative, which has helped more than 300 million young people intensively access computer science and entrepreneurial training.

The young women explained to me some of the projects they'd been working on. One team had decided to help some of the 115,000 refugees from war-torn Syria who had poured into Egypt since 2013. They'd built an app to help them find assistance upon their arrival in Egypt. But it was another group's project that really fascinated me. One group built an experience that digitally transforms the relationship between pharmacies and patients by making it easier, faster, and more affordable to find the pharmacy nearest you with the medications or supplies you need. Earlier in the day, I'd met an Egyptian entrepreneur who had built a similar app for locating the right doctor. In combination, they reminded me very much of Zocdoc, a New York City–based company that provides similar health-care services. Zocdoc became one of the celebrated American unicorns—nonpublic tech startup companies worth $1 billion or more. What I was witnessing firsthand was how quickly technology is diffused. Egyptian entrepreneurs were creating their own unicorns, even if they did not enjoy the lofty valuations of U.S.

startups. The core reason they were able to do so was because the cloud technology that enabled these innovations was now available to them without the need to invest a lot of capital.

Unfortunately, in many underserved parts of the world, public and private attention is focused on attracting Silicon Valley companies rather than on growing local tech entrepreneurs. Successful entrepreneurs in developing nations often tell me they can't even get meetings with their president or prime minister. Yet those same heads of state routinely meet with Western CEOs like me, looking for very near-term, foreign direct investment.

That's a shortsighted policy, and very frustrating for the business leaders who are trying to nurture the long-term prospects of their local and national economies. And yet I see this mindset everywhere—in the Middle East, Asia, Africa, Latin America, and even in struggling communities in G20 nations like the United States. The resulting failure of governments to encourage rapid and intensive use of new technologies means that the trend toward growing economic inequality between the haves and the have-nots of the world has continued unabated.

To get a measure of how equitable or inequitable our world is, economists turn to the work of an Italian economist named Corrado Gini who in 1912 published his formula for calculating what has become known as the Gini coefficient, which measures the difference between a society's division of income and a perfectly equal division of income. It's really quite elegant. If 100 percent of a given population were to earn $1 per day, that would be absolute equality. If 100 percent earned $1 million per year, that too would be absolute equality. But when only 1 percent earn

$1 million while everyone else earns nothing, we're approaching absolute inequality. Gini's work provides a way of measuring the degree to which the income distribution in a given society approaches or diverges from perfect equality.

The Gini coefficient for a particular population is generally expressed as a fraction. Perfect equality would be represented by a value of zero, while maximum inequality would be represented by a value of one. In the real world, the Gini coefficient for any given country or region is expressed by a fraction somewhere in between those two extremes. The Gini coefficient for an advanced European country like Germany has hovered around .3 for decades, while the figure for the United States has risen for years, now matching that of China and Mexico at over .4.

Of course, most economists agree that perfect income equality is neither possible nor desirable. Capitalist economies reward qualities like innovation, risk-taking, and hard work—qualities that generate value, produce wealth, and usually lead to benefits for many people throughout the society. When rewards flow to the people who exhibit those qualities, unequal income distribution is the inevitable result.

Edward Conard, a founding partner of Bain Capital, carries the argument even further in *The Upside of Inequality*. Conard concludes that inequality ultimately leads to faster growth and greater prosperity for everyone. Investors wait for good ideas that create their own demand for properly trained talent needed to commercialize ideas successfully. He sees two constraints on growth: an economy's capacity and willingness to take risk and to find properly trained and motivated talent.

But excessive inequality has the perverse effect of reducing incentives for many people. What happens when people work more and yet make less money? It is discouraging, leading many people to slacken their efforts, abandon dreams of launching or expanding businesses, and perhaps to leave the workforce altogether. It also weakens overall economic activity. For businesses like mine, it means our global customers have less to spend on emerging technologies that could make them more productive. That is what's happening today. There is a sagging line below Gini's perfect 45-degree angle that represents growing inequality. I want to avoid the pitfalls of what Marx described as late-stage capitalism—a theoretical time when economic growth and profits collapse—and get back to the returns enjoyed in early stage capitalism. But how? That's the question most heads of state around the world are also grappling with.

In computer science and engineering, we search for something called the *global maxima*. It's a mathematical phrase describing the optimal state—the highest point of a function. Where technology is concerned, I would argue that the global maxima for every region of the world—a country, county, or community—should be to import the latest world-class technologies in order to fuel innovation and growth among that nation or region's entrepreneurs—to drive both exports and local consumption of these innovations with intensity across sectors and segments of society. In other words, focus on adding value as well as broad use to help generate surplus and opportunity for more and more citizens. This means every region—in both developed and developing countries—must grow industries in which they have comparative economic advantage with use of new technology inputs. Business leaders and policymakers need to ask: What do we have that others do not have? And how can we turn that unique advantage into a source of growth and wealth for all our people?

China has clearly done this with proactive industrial policy that supports their entrepreneurs and economy across manufacturing and consumer Internet services. China strategically used the global supply chain and their own domestic market to amplify their comparative advantage and bootstrap their economic growth. The combination of industrial policy, public sector investment, and entrepreneurial energy is what many other countries will also look to replicate from China's success. I see the beginnings of this in India with the creation of the new digital ecosystem known as IndiaStack. India is leapfrogging from once being an infrastructure-poor country to now leading in digital technology. IndiaStack ushers in a presence-less, cashless, paperless economy for all its citizens.

On a trip to Bengaluru I engaged in a conversation with Nandan Nilekani about IndiaStack and its future road map. Nandan is the legendary founder of Infosys, who went on to create a new startup working with the Indian Government—Aadhaar—the identity system that is at the center of IndiaStack. Aadhaar now has scaled to over 1 billion people, rivaling the growth of other platform innovations such as Windows, Android, or Facebook.

Enlightiks, a startup that was acquired by Practo, is a leading e-health company in India. I met the founder of Enlightiks on the same trip to Bengaluru. They are using the latest cloud technology and AI from Microsoft to create a state-of-the-art health-care diagnostics service that can, for example, detect an atrial fibrillation event before it happens because of the rich data going from the personal device of the patient directly to the cloud. In turn, this cloud service can be made available to hospitals in smaller towns or rural areas in India. Enlightiks also has plans to take advantage of IndiaStack to authenticate the user, accept payment, create portal medical records, and much more. This Indian innovation is now looking to expand in the United States, Africa, and everywhere else.

This dynamic is not unique to China or India. I saw this across Chile, Indonesia, and Poland, and also in France, Germany, and Japan. Reflecting on my earlier visit to Egypt, it's clear they are investing in human capital. Egypt has an ancient heritage of science, math, and technology, and its universities have produced physicians who work throughout the Arab world. So health care turns out to be one of Egypt's areas of comparative advantage. The young entrepreneurs I met who have built apps for finding doctors and pharmacies are exploiting valuable synergies to create a powerful ecosystem, which is part of the magic of modern technology. Now they need affordable, powerful cloud services, which can come from Microsoft or another large-scale cloud provider. The right policy framework can help give their ideas flight.

Unfortunately, many governments have been resistant to embrace new technologies like the cloud even after they begin to reach scale in other parts of the world. In some cases, they try to pursue technology strategies that are self-defeating. For example, government leaders sometimes cite security, privacy, complexity, control, and latency (delayed processing) as reasons for building their own proprietary cloud rather than adopting an existing technology that has been made affordable by multinational demand.

Newly attuned to these issues and to the severe economic consequences they can produce, I returned from my Middle East trip with a renewed sense of energy and duty. I got off the airplane, marched into my office, and rallied our team to think through a set of recommendations and a policy framework to help governments, both developed and developing, reduce barriers to technology adoption and use.

So, back to the questions I posed earlier in this chapter. Are we growing, are we growing evenly, and what is the role of technology? There is, of course, no silver bullet, but as I consider all of the evidence and reflect on my own experiences, I keep returning to this simplified equation:

$$\sum (\text{Education} + \text{Innovation}) \times \text{Intensity of Tech Use} = \text{Economic Growth}$$

Education plus innovation, applied broadly across the economy and especially in sectors where the country or region has a comparative advantage, multiplied by the intense use of technology, over time, produces economic growth and productivity.

In a digital age, software acts as the universal input that can be produced in abundance and applied across both public and private sectors and every industry from agriculture to health care and manufacturing. Regardless of location—Detroit, Egypt, or Indonesia—this new input needs to turn into local economic surplus. Breakthrough technologies, plus a workforce trained to use them productively, multiplied by the intensity of their use spreads economic growth and opportunity. To make that happen, leaders need to prioritize entrepreneurship in several major ways.

The first is providing broad access to Internet connectivity and cloud computing services to all citizens. Today, such access is extremely variable. Internet penetration is close to 100 percent in Korea, Qatar, and Saudi Arabia, but below 2 percent in a number of sub-Saharan African nations. Unless we take specific steps to make access universal, by 2020 just 16 percent of people in the world's poorest countries and only 53 percent of the total global population will be connected to the Internet. At this rate, universal Internet access in low-income nations won't be achieved until 2042. And with no Internet access, there is no cloud access.

To expand Internet access, countries might adopt policies to facilitate the sharing of underutilized spectrum such as TV white space, an approach that is currently being successfully used in some developing countries. In addition, governments should lower restrictions on foreign direct investments in telecommunications, mobile, and broadband infrastructure, as well as reform other investment policies that erect barriers in the way of entrepreneurs willing to enter the market. Policies that encourage public-private partnerships and recognize the structuring needs of funding institutions are needed to facilitate access to capital for expanding Internet infrastructure.

Leaders at every level—from national to community-level—should foster not just fast but intense adoption of new technologies that can drive productivity. As Professor Comin told me, you don't have to invent the wheel, but you should adopt it quickly, because "societies that utilize new tools quickly are likely to be more productive."

Another high-priority area is fostering human capital and next-generation skills development. Building knowledge allows workers to keep up with the increasing pace of technology. As the digital transformation automates many tasks formerly handled by people, workers need the skills that will enable them to become managers of the new automated tools. Just as workers wielding shovels gave way to workers capable of driving bulldozers, societies now need people with the skills to manage fleets of automated bulldozers, self-driving cars, and drones.

To this end, government must demonstrate empathy for all of its constituents, and work to create a more knowledge-based economy. The pathway to new technologies requires a parallel investment in skills development—making sure people have the requisite skills to participate in an increasingly digital society, one that depends on smart devices and online services. In schools, this requires promoting digital literacy and making sure that teachers and students have access to technology and learning tools at low cost. In the workplace, we need to invest in lifelong learning with a focus on programs and investments that promote upskilling for the cloud and a more digital-ready workforce. Companies like Microsoft are already expanding their educational capacity and building initiatives to accelerate such skills development, especially at small- and medium-sized enterprises.

Knowledge is necessary to find new uses for new technologies and that knowledge is accumulated through training and experience. Every country is different, but Germany provides an excellent example of the productive use of new technologies. Germany and the United States both invest heavily in R&D, but Germany has enjoyed greater rates of productivity growth. Why? One explanation is the German system of vocational training through apprenticeship, which makes cutting-edge technologies available to the workforce quickly through vocational schools that have close relationships with industry. I am convinced the only way to tackle economic displacement is to make sure that we provide skills training not only to people coming out of college and other postsecondary programs, but also to workers who are losing their jobs to automation. Countries that invest in building technology skills as a percent of GDP will see the rewards.

Policy reforms must also create a regulatory environment that promotes innovative and confident adoption and use of technology. While data privacy and security are always key concerns, they also need to be balanced against the demands for data to flow more freely across borders and between the various services that make up a modern global digital economy. Governments have been strong advocates for promoting digital security to protect the community from harm. However, our experience is that public policy and regulation in this area needs reform to ensure the right balance is struck. This is by no means easy, but Microsoft and other leaders in our industry have extensive experience helping governments modernize their regulatory frameworks to achieve this balance and help promote public safety and national security without compromising the benefits of these digital services for the public and private sectors and millions of citizens.

Additionally, every government has an opportunity to lead by example in embracing technology for the provision of services to citizens, improving productivity in the public sector, and leveraging its comparative advantage. Public sector leadership should be complemented by efforts to showcase local entrepreneurship and leading-edge technologies, including providing financial incentives where appropriate.

As leaders ask themselves, "Where can we be the best in the world?" the answers might be surprising—desert farming in Australia or local banking in Dubai. Some other country or community might strive to become the world's leader for innovations in IoT; ambient intelligence; mobile payment systems; virtual reality; silicon photonics; 3D printing; wearables; lightweight, low altitude satellites; drones; native advertising; driverless cars; robotics and industrial automation; adaptive, gamified education; nano-machines; genomics; or economical solar, wind, and tidal power. Each represents an opportunity for leadership that no single community or region has yet seized. Seattle, for example, has become the center of excellence for cloud computing as the home of both Amazon and Microsoft.

An inspiring idea in this context is the notion of a charter or startup city, an idea put forth by economist Paul Romer. Romer posits that rules and laws, which are hard to change and require concessions to be approved, are not optimized for spurring innovation and creating economic growth. Charter cities, on the other hand, are experimental reform zones engineered entirely to create jobs and growth. Citizens could opt in or not. Some will be ready and some won't. His illustration is Hong Kong and Shenzhen. Hong Kong, located in China but ruled for generations by Great Britain, was free of antimarket Communist rule and became an economic engine, attracting and training workers. Deng Xiaoping, grasping that China needed to become more open in order to grow, created a de facto charter city in nearby Shenzhen, which could take advantage of its neighbor's talent pool and infrastructure. Unlike the rest of China, Shenzhen's rules would be attractive to foreign investment and international trade. He knew that Communist China would be slow to embrace these reform zones, but many entrepreneurs and workers would leap at the opportunity. Shenzhen grew from a town of thirty thousand people to a global financial center of nearly 11 million residents after it was designated as a special economic zone in 1980.

We also need to continue to promote free and fair trade. If we want to see growth and see it more broadly, opening up more markets and clearing barriers to trade for entrepreneurs is an essential step. It's unfortunate that, in recent years, populist politicians on both the left and the right have campaigned on pledges to overturn free-trade agreements.

Governor John Kasich of Ohio, in the midst of a bitter 2016 presidential campaign, wrote an op-ed for *The Wall Street Journal* arguing that a vote against trade is a vote against growth. He pointed out that the Trans-Pacific Partnership (TPP), a major trade agreement that was pending approval in Washington at the time, is about helping large and small companies find growth in Japan, Australia, Canada, Chile, and other Pacific Rim nations that want to increase trade with America. The world needs continued progress on trade liberalization. Kasich pointed out that 40 million U.S. jobs depend on trade. But our trade laws also need to be modernized. In this digital economy, the bits and bytes imported and exported are as critical to trade as the automobiles, agricultural products, and other goods we trade. We need to be able to transfer data across borders in the course of business and to do so without having to locate computing facilities in every territory, while protecting privacy, source code, and other forms of intellectual property.

The 2016 campaigns brought new attention to the challenges and benefits of trade agreements. Despite a lot of noise, all of the candidates said that they thought trade was good, but each had differing views. Trump on the right and Sanders on the left suggested massive job loss. Clinton focused on the need for stronger enforcement. Business leaders argued that trade deals have been net job creators, though I am sympathetic with the view that those gains need to be distributed more evenly. To complaints that trade agreements are bad for the environment, supporters have pointed out that TPP is the first multilateral trade deal to include enforceable provisions for environmental protection.

To be sure, the basic rules-based framework for economic relations between nations that were established at the end of

World War II through the Bretton-Woods system provides a good, but imperfect, foundation on which to build. The framework continues to form the foundational principles for closer cooperation with like-minded countries through networks of free-trade agreements. But, trade agreements will only continue to be successful if they are seen in the broader context of economic policies for growth.

Finally, questions are being asked about whether this next industrial revolution will be a jobless one. To help us investigate this question, MIT economist Daron Acemoglu visited our campus to report on his research into the effects of technology automation on labor. He found that new intelligent machines, particularly industrial robots, could have very consequential effects on the labor market. His estimates suggest that, on average, each additional industrial robot reduces employment by about three workers. This suggests that, without any countervailing changes, the spread of industrial robots could have very adverse consequences for jobs and wages. Nevertheless, Acemoglu argues that other powerful changes triggered by this onslaught could at least partly reverse these consequences. As machines replace labor in some tasks, firms will be incentivized to create new tasks in which humans have a comparative advantage. Acemoglu sums it up this way: "Although automation tends to reduce employment and the share of labor in national income, the creation of more complex tasks has the opposite effect." Throughout history new classes of workers and new, more complex tasks have resulted from cutting-edge technologies. Acemoglu continues, "The creation of new complex tasks always increases wages, employment and share of labor. But when automation runs ahead of the process of creation of new, labor-intensive tasks, technological change will bring lower employment." We need a balanced growth path. We need to invent a new social contract for this age of AI and automation that fosters the equilibrium between individual labor—one's agency, wages, sense of purpose, and fulfillment—and the return on capital.

One such example is Kent International, maker of Bicycle Corporation of America–branded bikes, which made news early in 2017 when it moved 140 jobs from China back to Manning, South Carolina, where the company had invested in robotics to automate many of the tasks once performed by people. What was once a low-tech, high-labor business is going through its own digital transformation and plans. CEO Arnold Kamler told me he plans to add forty jobs per year, which is considerable growth in a small town. In fact, a number of states competed to land the plant. "A lot of people have that misconception that automation decreases jobs," a production manager on the line said. "It's just a different type of job, a more skilled job." Without the robots, the human jobs wouldn't exist.

One of the reasons I was so excited about our acquisition of LinkedIn—the talent and employment-oriented social networking service—was a shared commitment we discovered early in our negotiations. In conversations with LinkedIn's founder, Reid Hoffman, and its CEO, Jeff Weiner, I discovered we had a common desire to use our digital platforms to spread opportunity more equitably to everyone. In fact, *The New Yorker* magazine has written about LinkedIn's vision of making labor markets work better for all 3 billion members of the global workforce by making those markets more efficient and open.

This dream of a more accessible and equitable economic playing field won't come true automatically. In his book *The Start-up of You*, Hoffman writes about the forces of competition and change that brought down Detroit as an economic powerhouse: "No matter what city you live in, no matter what business or industry you work for, no matter what kind of work you do—when it comes to your career, right now, you may be heading down the same path as Detroit." What we aspire to do through LinkedIn is to build a network of alliances to help provide the intelligence on opportunities, training resources, and collective action we all can take to create economic opportunity for individuals. In this way, we hope to ensure that other cities in the future won't suffer the same fate as Detroit—and, in fact, just as Detroit did, other cities can shape their own successful reinventions as thriving centers of enterprise and job creation for decades to come.

I do have a bias for which I am unapologetic. It is a bias for driving investment toward technological advancements in services like LinkedIn and Office that help people create, connect, and become more productive rather than software that is simply entertaining—memes for conspicuous consumption. Spillover effects on the economy are pretty limited for technologies that don't foster a more equitable ratio of consumption to creation. Nonetheless, Wall Street has put a lot of value recently on these consumption technologies.

Robert Gordon's recent economic treatise, *The Rise and Fall of American Growth*, has as its central thesis that some inventions are more important than others. I agree, and I would put today's productivity software in that category. Gordon examines American growth between 1870 and 1940, describing a century of economic revolution that freed households from an unremitting daily grind of painful manual labor, household drudgery, darkness, isolation, and early death. It was a transformation unique in human history, unrepeatable because so many of its achievements could happen only once. Looking over the great expanse of American economic history, Gordon concludes that *innovation* is the ultimate source of such dramatic changes: "Entrepreneurs contribute to economic growth far more than the narrow word 'innovation' can convey," he writes. And education, he further notes, is innovation's closest cousin in fueling growth.

John Batelle, *Wired*'s co–founding editor, once wrote that "Business is humanity's most resilient, iterative, and productive mechanism for creating change in the world." He is right—and we business leaders need to take seriously our responsibilities as change leaders. I don't say this for purposes of so-called corporate social responsibility, which is important but can also serve as little more than good PR. I say it because a better world is better for business. It's important to be dedicated to creating great products, serving customers, and earning profits for our investors—but it's not sufficient. We also need to think about the impact of our actions on the world and its citizens long into the future.

Afterword

"Why do I exist?"

"Why does our institution exist?"

"What is the role of a multinational corporation in our world?"

"What is the role of a leader in digital technology, especially as the world turns to tech as such a crucial input to drive growth?"

These questions haunt me, and they motivated me to write *Hit Refresh*. Finding answers set me on an intellectual and introspective journey to discover what I uniquely can contribute to society and how to rediscover the soul of Microsoft, to define our role as a global company. They guide me daily in the pursuit to bring empathy together with big ideas in order to make a real difference. Hopefully the stories and lessons along the course of my journey have produced something useful to you in your own life and work.

I also hope these existential questions spark conversations among policymakers, business leaders, and technologists. In an often-divided world that is careening toward ever more dramatic technological, economic, demographic, and even climactic shifts—we have to redefine the role of multinational corporations and the role of leadership. Anti-globalization movements like Brexit and populist political campaigns in both America and

Europe have raised important questions and concerns ranging from automation, trade, and economic opportunity, to fairness and whom to trust.

Economist Richard Baldwin, author of *The Great Convergence*, writes that the origin of today's anti-globalization sentiment in the wealthiest nations lies in the fact that their share of world income has plummeted from 70 percent in 1990 to 46 percent in just the past two decades. In other words, wealthy nations like the United States, France, Germany, and the UK have witnessed a large drop in their share of world income. The combination of low wages and information technologies that radically lowered the cost of moving ideas has meant that places like China and India have significantly gained in the share of world income while rich nations are now back to 1914 levels, igniting anti-globalization feelings in some quarters. Baldwin predicts a third wave of globalization will come when telepresence and telerobotics (like HoloLens)—really good substitutes for people crossing borders to provide services—become affordable.

As this book was going to press, Nobel economist Angus Deaton and his wife, Anne Case, also a distinguished economist at Princeton, published a paper that found whites in the U.S. who have less than a college degree, experience cumulative disadvantages over the course of their lives that can negatively impact their mortality, health, and economic well-being. In fact, their research found that it is education more than income that explains increases in mortality and morbidity among whites in midlife. This dynamic, coupled with Baldwin's findings, have at least in part fueled today's anti-globalist fervor, and as a result now invite introspection on both our public education and public health priorities.

Of course, the goal is to grow the pie for everyone. GE's Jeff Immelt provided his answer to the role of today's multinational in a 2016 speech to the Stern School of Business at New York University. He reflected on the role played by global companies during his thirty-year career. During that time, the rate of extreme poverty had been cut in half, and technological innovations had dramatically improved health care, reduced the cost of energy, and connected people like never before. Yet today, Immelt observed, big companies are seen as failing (along with governments) to address the world's challenges. In response, Immelt announced that GE intends to pivot. Because greater equality is good both for business and for society, GE plans to adopt policies designed to help level the playing field globally. GE will localize—that is, it will grow local capabilities within the company's global footprint, making space for greater local decision-making in a more comprehensive local context.

I agree. More than half of Microsoft's revenues are from outside the United States. We can't do business effectively in 190 countries unless we prioritize the creation of greater local economic opportunity in each of those countries. We've invested more than $15 billion in constructing thirty of the world's most sophisticated regional data centers, positioning them to support local entrepreneurship and public sector services in North America, South America, Asia, Africa, and Europe. In each of these regions, we have to operate with responsibility. Real business success, in fact capitalism generally, cannot be just the surplus that you create for your own core constituency, but also the broader surplus that is created to benefit the wider society.

The way I look at it, multinationals can no longer be the memes they've become—soulless, bloodless entities that enter a nation or a region simply to take rent from the locals. The job of a multinational is more important than ever. It needs to operate everywhere in the world, contributing to local communities in positive ways—sparking growth, competitiveness, and opportunity for all. How can we help our local partners and startups grow? How can we help the public sector become more efficient? How can we help solve the most pressing issues in society, like access to education and health? Every country, naturally, thinks about its own national interest first. America first in America. India first in India. The UK first in the UK. The priority of a global company should be to operate in each of these countries with the goal of creating local opportunity in long-term, sustainable ways.

We all must do this while staying steadfastly anchored in our timeless values. Microsoft is a company born in America. Our heritage has shaped our values. We believe in the American Dream—both in living it out as employees and helping others do the same. Our allegiance is to a set of enduring values—privacy, security, free speech, opportunity, diversity, and inclusion. We live by them, and we will stand for them when challenged in America and elsewhere.

Multinational corporations that create technology have an even higher bar for creating economic opportunity as the next wave of technologies takes hold. The coming industrial revolution, one that builds toward ubiquitous computing and ambient intelligence and is fueled by software, will be more profound in its impact on the economy than those revolutions that came before. It's why I developed a set of design principles that shape how we—Microsoft and others—create this next wave of technology. I encourage feedback, debate, and ultimately commitment to building out the ethics that will govern our society going forward.

The world used to grow at 4 percent per year, but it is now growing at roughly 2 percent. So, we need new technology breakthroughs in order to have the type of growth we had in the twentieth century. Mixed reality, artificial intelligence, and quantum computing are going to be game changers, creating new economic surplus, but also disrupting the workforce, eliminating the routine jobs we take for granted today. Some argue that robots will take all our jobs, but this so-called "lump of labor" argument—the notion that there is a limited amount of work available—has always been disproved. It's just that different types of labor will be needed. And humans will add value where machines cannot. As we encounter more and more artificial intelligence, real intelligence, real empathy, and real common sense will be scarce. The new jobs will be predicated on knowing how to work with machines, but also on these uniquely human attributes.

In the face of these many coming shifts, there must be a new social contract that helps to achieve economic surplus and opportunity on a more equitable basis. To get there, what will the new labor movement look like? There has been talk of a Universal Basic Income. How will we re-skill and retrain workers—not just high-end knowledge workers, but also low-skill and mid-skill labor? Can the service sector and people-on-people jobs be the source of new employment for many displaced from traditional manufacturing or agricultural sectors?

Finally, as leaders, what is our role? At the end of the day, leaders of any company are evaluated based on their ability to grow the business, to clear the way for innovations that inspire customers. As CEOs, we're accountable for generating the best returns to shareholders. But I also subscribe to the notion that the bigger a company is, the more responsibility its leader has to think about the world, its citizens, and their long-term opportunities. You're not going to have much of a stable business if you don't think about the growing inequities around the world and do your part to help improve conditions for everyone.

We approach this goal by focusing on multiple strategies and constituencies, leveraging our core business for positive social impact, and improving personal productivity, making sure our business is socially responsible by investing in sustainability, accessibility, privacy and security, and through philanthropy. Microsoft Philanthropies is the world's largest corporate philanthropy with more than $1 billion in contributions across a wide variety of causes including teaching digital skills like coding and computer science, affordable access to the Internet, and humanitarian assistance. And we use our voice—under the banner of A Cloud for Global Good—to advocate for policies that advance the goal of economic opportunity for everyone. In fact, all of my proceeds from this book will go toward these causes.

Earlier I wrote that the C in CEO is about being the curator of culture. After all, it really comes down to people. It's the sum of a million decisions made by thousands of people every day. It's about helping employees live out their own personal mission in the context of Microsoft's. Microsoft no longer employs people, people employ Microsoft. What is possible to achieve when we shift the mindset of more than 100,000 people from being employees to employers? Our entire purpose is to make things that help others make things—and make things happen. Our services are irreplaceable tools in countless enterprises and organizations around the world. Anybody at Microsoft can look at our constellation of assets and dream of what can be and bring it to bear on any problem in any geography. We're providing the resources countless people can use to build something that will outlast themselves, whether that's a small business, a school, a clinic, or a giant enterprise creating jobs and opportunity for millions.

This culture needs to be a microcosm of the world we hope to create outside the company. One where builders, makers, and creators achieve great things. But, equally important, one where every individual can be their best self, where diversity of skin color, gender, religion, and sexual orientation is understood and celebrated. I know that we are on the right track when I hear a colleague express an insight that could only come from empathy, or when a product breakthrough results from someone who used Microsoft as a platform for his or her personal passion and creativity.

What does it mean to *Hit Refresh*? I encourage you to answer that for yourself. Start the conversation in your institution. Start the conversation in your community. And please, share with me what you learn, and I'll continue to do the same.

Acknowledgments

I've often said that the best lines of computer code are like poetry. The writer struggles to compress so much thought and feeling into the fewest lines possible while still communicating the fullness of expression. Although the prose we've written here does not approach poetry, the writing process was nonetheless intense, and in the end rewarding. For that, there are many people to thank.

As I wrote in my dedication, I owe so much to two families. At home, Anu and our three beautiful children as well as our parents back in India.

My other family has been Microsoft for more than two decades. I owe a lot to Bill Gates, Paul Allen, and Steve Ballmer, who together created the opportunity for all of us at Microsoft to innovate, scale, and serve customers around the world. I have admired and learned from each of them throughout my career. Our senior leadership team is my partner in this continuing transformation, and I want to offer them my most sincere thanks and appreciation: Judson Althoff, Chris Capossela, Jean-Philippe Courtois, Kurt DelBene, Scott Guthrie, Kathleen Hogan, Amy Hood, Rajesh Jha, Peggy Johnson, Terry Myerson, Kevin Scott, Harry Shum, Brad Smith, and Jeff Weiner. None of our work would be possible without the creativity and talent of every single Microsoft employee and partner.

Our board of directors: John Thompson, Reid Hoffman, Teri L. List-Stoll, G. Mason Morfit, Charles H. Noski, Dr. Helmut Panke, Sandra E. Peterson, Charles W. Scharf, John W. Stanton, and Padmasree Warrior.

My coauthors and I relied on a number of experienced publishing hands from start to finish. Karl Weber brought a gifted touch both to the development of the manuscript and to the copy itself. Jim Levine, my agent, was always a calm, guiding voice throughout the process. And our editor and publisher, Hollis Heimbouch at HarperCollins, was encouraging before we had even written a word, took a chance on our ideas once we jotted them down, and served as our Virgil through the dark forest.

The Microsoft Library and Archives team, Kimberly Engelkes, Nicole Partridge, and Amy Stevenson, provided invaluable fact-checking and a useful set of notes at the end of the book.

I cannot thank my terrific staff enough for their daily support—Jason Graefe, Cynthia Thomsen, Bonita Armstrong, Caitlin McCabe, Colette Stallbaumer, Chad DeVries, Megan Gray, Jeff Furey, and the entire team.

Our communications and marketing experts, including Frank X. Shaw, Bob Bejan, Steve Clayton, Doug Dawson, and John Cirone and their teams. This group was invaluable in reading the manuscript, partnering with HarperCollins, and getting the word out about the book.

Special thanks to Matthew Penarczyk in our legal department, and to the many who contributed ideas and thinking throughout:

Acknowledgments

Rolf Harms, Jon Tinter, Matt Booty, Alex Kipman, R. Preston McAfee, Justin Rao, Glen Weyl, Victor Heymeyer, Mike Tholfsen, Nate Jones, Turi Widsteen, Chinar Bopshetty, Michael Friedman, Krysta Svore, Peter Lee, Eric Horvitz, Kate Crawford, Danah Boyd, Chris Bishop, Dev Stahlkopf, John Seethoff, Abigail Sellen, Ryan Calo, and Prem Pahlajrai. Sports journalist Suresh Menon, editor of *Wisden India Almanack*, suggested the cricket writing used in Chapter 2, and was kind enough to offer helpful guidance.

Walter Isaacson not only provided early input for the book's direction but also interviewed me onstage at the Aspen Ideas Festival where we first announced the book. Tina Brown and her husband, Harold Evans, kindly hosted Anu and me at their wonderful home in New York City where we discussed Microsoft and some of the ideas in the book with other writers and thinkers. Tim O'Reilly interviewed me on these topics at his innovative What's the Future (WTF) conference in San Francisco, and I wish him the best of luck with his latest book.

Lastly, I want to thank Greg Shaw and Jill Tracie Nichols, my coauthors, for their partnership on this project—for encouraging me to pursue it, for helping me to craft it, and for working with me to make it as meaningful as possible.

Sources and Further Reading

CHAPTER 1 — FROM HYDERABAD TO REDMOND

Cornet, Manu. "Organizational Charts." Bonkers World, June 27, 2011. Accessed December 8, 2016. http://www.bonkersworld.net /organizational-charts/.

Gordon, Robert J. *The Rise and Fall of American Growth: The U.S. Standard of Living since the Civil War.* Princeton, NJ: Princeton University Press, 2016.

Widmer, Ted. "The Immigration Dividend." *New York Times*, October 6, 2015.

CHAPTER 2 — LEARNING TO LEAD

A Cloud for Global Good. Case study. Redmond, WA: Microsoft, 2016. Accessed December 12, 2016. http://news.microsoft.com /cloudforgood/.

Guha, Ramachandra. *A Corner of a Foreign Field: The Indian History of a British Sport.* Basingstoke, UK: Pan Macmillan, 2003.

Eastaway, Robert. *Cricket Explained.* New York: St. Martin's Griffin, 1993.

Shapshak, Toby. "How Kenya's M-Kopa Brings Prepaid Solar Power To Rural Africa." *Forbes*, January 28, 2016.

Beser, Ari. "How Citizen Science Changed the Way Fukushima Radiation is Reported." *National Geographic*, Fulbright National Geographic Stories, February 13, 2016.

Heikell, Lorence. "UN and Microsoft Aid Disaster Recovery, Economic Development in Nepal." Microsoft Feature Story. Accessed March 10, 2017. https://news.microsoft.com/features /un-and-microsoft-aid-disaster-recovery-economic-development -in-nepal/#sm.00001tfvv5hhqcs610r97vxf4vfiv#hAyXgOep0YzF R1W8.97.

Amazon. "New Version of Alexa Web Search Service Gives Any Developer Tools to Innovate in Search at Web Scale." Amazon Press Release, June 6, 2007. http://phx.corporate-ir.net/phoenix .zhtml?c=176060&p=irol-newsArticle&ID=1012591.

Barr, Allison. "Amazon's Next Billion-dollar Business Eyed." Reuters, July 22, 2011.

Brengel, Kellogg. "ThyssenKrupp Elevator Uses Microsoft Azure IoT for Improved Building Efficiency." OnMicrosoft. Accessed March 10, 2017. https://www.onmsft.com/news/thyssenkrupp-elevator -uses-microsoft-azure-iot-improved-building-efficiency.

CHAPTER 3 — NEW MISSION, NEW MOMENTUM

Vance, Ashlee. "CEO Memo Makes 'Productivity' the New Mantra at Microsoft." *Bloomberg*, July 10, 2014.

McGregor, Jen. "Microsoft CEO Satya Nadella's Love of Literary Quotes." *Washington Post*, July 10, 2014.

Wingfield, Nick. "Satya Nadella Says Changes Are Coming to Microsoft." *New York Times*, July 10, 2014.

CHAPTER 4 — A CULTURAL RENAISSANCE

Peckham, Matt. "'Minecraft' Is Now the Second Best-Selling Game of All Time." *Time*, June 2, 2016.

CHAPTER 5 — FRIENDS OR FRENEMIES?

http://spectrum.ieee.org/tech-talk/telecom/internet/popular -internet-of-things-forecast-of-50-billion-devices-by-2020-is -outdated.

CHAPTER 6 — BEYOND THE CLOUD

Linn, Allison. "How Microsoft Computer Scientists and Researchers Are Working to 'Solve' Cancer." Microsoft Story Labs, September 2016. https://news.microsoft.com/stories/computingcancer/.

Dupzyk, Kevin. "I Saw the Future Through Microsoft's HoloLens." *Popular Mechanics,* September 6, 2016, http://www.popularmechanics.com/technology/a22384/hololens-ar-breakthrough-awards/.

Aukstakalnis, Steve. *Practical Augmented Reality. A Guide to the Technologies, Applications, and Human Factors for AR and VR.* Boston: Addison-Wesley, 2016.

Grunwald, Martin. *Human Haptic Perception: Basics and Applications.* Boston: Birkhauser, 2008.

Gartner, Hype Cycle for Emerging Technologies, 2016, G00299893

Aaronson, Scott. *Quantum Computing Since Democritus.* Cambridge: Cambridge University Press, 2013.

Linn, Allison. "Microsoft Doubles Down on Quantum Computing Bet." Next at Microsoft Blog, November 20, 2016. https://blogs.microsoft.com/next/2016/11/20/microsoft-doubles-quantum-computing-bet/.

CHAPTER 7 — THE TRUST EQUATION

Ignatius, Adi. "They Burned the House Down." *Harvard Business Review* 93, no. 7/8 (2015): 106–13.

Smith, Brad. "'The Interview' Now Available on Xbox Video." The Official Microsoft Blog, December 24, 2014. http://blogs.microsoft.com/blog/2014/12/24/the-interview-now-available-on-xbox-video/.

Microsoft News Center. "Statement from Microsoft about Response to Government Demands for Customer Data." The Official Microsoft Blog, July 11, 2013. http://news.microsoft.com/2013/07/11/statement-from-microsoft-about-response-to-government-demands-for-customer-data/#sm.001aorusr7vufs511ur2bludrw2u3.

Hesseldahl, Arik. "Microsoft and Google Will Sue U.S. Government Over FISA Order Data." *All Things D*, August 30, 2013. http://allthingsd.com/20130830/microsoft-and-google-will-sue-u-s-government-over-fisa-order-data/#.

Cellan-Jones, Rory. "Technology Firms Seek Government Surveillance Reform." *BBC Technology News*, December 9, 2013. Accessed December 8, 2016. http://www.bbc.com/news/technology-25297044.

Ackerman, Spencer. "Tech Giants Reach White House Deal on NSA Surveillance of Customer Data." *The Guardian*, January 27, 2014. Accessed December 8, 2016. https://www.theguardian.com/world/2014/jan/27/tech-giants-white-house-deal-surveillance-customer-data.

Ellingsen, Nora. "The Microsoft Ireland Case: A Brief Summary," LawFare Blog, July 15, 2016, https://www.lawfareblog.com/microsoft-ireland-case-brief-summary.

Bennet, James, et al. "Adapting Old Laws to New Technologies; Must Microsoft Turn Over Emails on Irish Servers?" *New York Times*, July 27, 2014. http://www.nytimes.com/2014/07/28/opinion/Must-Microsoft-Turn-Over-Emails-on-Irish-Servers.html?_r=0.

Conger, Kate. "The Federal District Court Ruled in Favor of U.S. Prosecutors, but We Appealed the Decision, and the United States Court of Appeals for the Second Circuit Backed Microsoft's Position." *TechCrunch*, July 14, 2016. https://techcrunch.com/2016/07/14/microsoft-wins-second-circuit-warrant/.

Nakashima, Ellen. "Apple Vows to Resist FBI Demand to Crack iPhone Linked to San Bernardino Attacks." *Washington Post*, February 17, 2016. Accessed December 8, 2016. https://www.washingtonpost.com/world/national-security/us-wants-apple-to-help-unlock-iphone-used-by-san-bernardino-shooter/2016/02/16/69b903ee-d4d9-11e5-9823-02b905009f99_story.html.

Bloomberg, Michael. "The Terrorism Fight Needs Silicon Valley; Tech Executives Are Dangerously Wrong in Resisting the Government's Requests for Their Help." *Wall Street Journal*, June 29, 2016.

Accessed December 8, 2016. http://www.wsj.com/articles/the
-terrorism-fight-needs-silicon-valley-1467239710.

Hazelwood, Charles. "Trusting the Ensemble." TED Talk, 19:36,
filmed July 2011. http://www.ted.com/talks/charles_hazlewood.

Gates, Bill. "Memo from Bill Gates." The Official Microsoft Blog,
January 11, 2012. http://news.microsoft.com/2012/01/11/memo
-from-bill-gates/#sm.00000196kro2y0ndaxxlau37xidty.

Delgado, Rick. "A Timeline of Big Data Analytics." *CTO Vision*, September
12, 2016. https://ctovision.com/timeline-big-data-analytics/.

Lieberman, Mark. "Zettascale Linguistics." *Language Log*, November
5, 2003. http://itre.cis.upenn.edu/~myl/languagelog
/archives/000087.html.

North, Douglass Cecil. *Economic Growth of the United States, 1790–
1860*. Englewood Cliffs, NJ: Prentice Hall, 1961.

Adams, John. "John Adams to Abigail Adams, 3 July 1776." Adams
Family Papers: An Electronic Archive, Massachusetts Historical
Society, Boston. Accessed December 8, 2016. http://www
.masshist.org/digitaladams/archive/doc?id=L17760703jasecond.

Riley v. California, 134 S. Ct. 2473, 189 L. Ed. 2d 430, 2014 U.S. LEXIS
4497, 82 U.S.L.W. 4558, 42 Media L. Rep. 1925, 24 Fla. L. Weekly
Fed. S 921, 60 Comm. Reg. (P & F) 1175, 2014 WL 2864483 (U.S.
2014). https://www.supremecourt.gov/opinions/13pdf/13
-132_8l9c.pdf.

Rothman, Lily. "10 Questions with Akhil Reed Amar." *Time*,
September 5, 2016, 56.

Arun K. Thiruvengadam Scholarly Papers. New York: Social Science
Research Network, 2013–2016. Accessed December 8, 2016.
https://papers.ssrn.com/sol3/cf_dev/AbsByAuth.cfm?per_
id=411428.

Malden, Mary, and Lee Rainie. "Americans' Attitudes about Privacy,
Security and Surveillance." Washington, DC: Pew Research Center,
2015. Accessed December 8, 2016.

http://www.pewinternet.org/files/2015/05/Privacy-and-Security
-Attitudes-5.19.15_FINAL.pdf.

Neuborne, Burt. *Madison's Music: On Reading the First Amendment.* New York: The New Press, 2015.

CHAPTER 8 — THE FUTURE OF HUMANS AND MACHINES

Markoff, John, and Paul Mozur, "For Sympathetic Ear, More Chinese Turn to Smartphone Program." *New York Times*, July 31, 2015.

Tractica. Virtual Digital Assistants. Boulder, CO: Tractica, 2016. Accessed December 8, 2016. https://www.tractica.com/research/virtual-digital-assistants/.

Executive Office of the President National Science and Technology County Committee on Technology. *Preparing for the Future of Artificial Intelligence.* Washington, DC: National Science and Technology Council, 2016. Accessed December 8, 2016. https://www.whitehouse.gov/sites/default/files/whitehouse_files/microsites/ostp/NSTC/preparing_for_the_future_of_ai.pdf.

Kurzweil, Ray. *The Singularity Is Near: When Humans Transcend Biology.* New York: Penguin Books, 2006.

Markoff, John. *Machines of Loving Grace: The Quest for Common Ground Between Humans and Robots.* New York: Ecco, 2015.

Asimov, Isaac. "Runaround." In *I, Robot.* New York: Gnome Press, 1950.

Gates, Bill. "The Internet Tidal Wave." Memorandum to executive staff, May 26, 1995. https://www.justice.gov/sites/default/files/atr/legacy/2006/03/03/20.pdf.

Breazeal, Cynthia. *Designing Sociable Robots.* London: MIT Press, 2002.

Nadella, Satya. "The Partnership of the Future." *Slate*, June 28, 2016. Accessed December 8, 2016. http://www.slate.com/authors.satya_nadella.html.

Stone, Peter, et al. "Artificial Intelligence and Life in 2030." *One Hundred Year Study on Artificial Intelligence: Report of the 2015–2016 Study Panel.* Stanford, CA: Stanford University, 2016. Accessed: September 6, 2016. https://ai100.stanford.edu/2016-report/preface.

Allen, Colin. "The Future of Moral Machines." *New York Times,* December 25, 2011.

Bostrom, Nick. *Superintelligence: Paths, Dangers, Strategies.* Oxford: Oxford University Press, 2014.

Ford, Martin. *Rise of the Robots: Technology and the Threat of a Jobless Future.* New York: Basic Books, 2015.

Brynjolfsson, Erik, and Andrew McAfee. *The Second Machine Age: Work, Progress, and Prosperity in a Time of Brilliant Technologies.* New York: W. W. Norton, 2014.

McCullough, David. *The Wright Brothers.* New York: Simon & Schuster, 2015.

Krznaric, Roman. *Empathy: Why It Matters, and How to Get It.* New York: TarcherPerigee, 2014.

Schwab, Klaus. *The Fourth Industrial Revolution.* New York: Crown Business, 2017.

Susskind, Daniel, and Richard Susskind. *The Future of the Professions: How Technology Will Transform the Work of Human Experts.* Oxford: Oxford University Press, 2016.

CHAPTER 9 — RESTORING ECONOMIC GROWTH FOR EVERYONE

Associated Press. "Who's Been Invited to the State of the Union Tonight?" *Boston Globe,* January 12, 2016. Accessed December 9 2016. https://www.bostonglobe.com/news/politics/2016/01/12/guestsrdp/DR3KzNA90x3nxLYFOFs0nN/story.html.

Obama, Barack. State of the Union Address. White House, January 12, 2016. Accessed December 9, 2016. https://www.whitehouse.gov/sotu.

Solow, Robert M. "We'd Better Watch Out." Review of *The Myth of the Post-Industrial Economy,* by Stephen S. Cohen and John Zysman. *New York Times,* July 12, 1987. Accessed December 9, 2016. http://www.standupeconomist.com/pdf/misc/solow-computer-productivity.pdf.

Nadella, Satya, Ulrich Spiesshofer, and Andrew McAfee. "Producing Digital Gains at Davos." *BCG Perspectives,* March 9, 2016. Accessed

December 9, 2016. https://www.bcgperspectives.com/content
/articles/technology-digital-technology-business-transformation
-producing-digital-gains-davos/.

Weightman, Gavin. *The Industrial Revolutionaries: The Making of the
Modern World, 1776–1914*. New York: Grove Press, 2010.

Ashton, T. S., and Pat Hudson. *The Industrial Revolution, 1760–1830*,
2nd ed. Oxford: Oxford University Press, 1998.

Republic of Malawi. National ICT Policy. Lilongwe: Malawi, 2013.
Accessed December 9, 2016. https://www.malawi.gov.mw
/Publications/Malawi_2013_Malawi_ICT_Policy.pdf.

Republic of Rwanda Ministry of Finance and Economic Planning.
Rwanda Vision 2020. Kigali: Rwanda, 2000. Accessed December 9,
2016. http://www.sida.se/globalassets/global/countries-and
-regions/africa/rwnda/d402331a.pdf.

Comin, Diego A., and Bart Hobijn. "Historical Cross-Country
Technology Adoption (HCCTA) Dataset." The National Bureau of
Economic Research. Last modified August 8, 2004. http://www
.nber.org/hccta/.

McKenzie, David, and Christopher Woodruff. "What Are We
Learning from Business Training and Entrepreneurship
Evaluations around the Developing World?" Working Paper
WPS6202, The World Bank Development Research Group Finance
and Private Sector Development Team. World Bank, 2012.
http://documents.worldbank.org/curated/en/777091468331811120
/pdf/wps6202.pdf.

Adesanya, Ireti. "The Genius Behind the Gini Index." Virginia
Commonwealth University School of Mass Communications
Multimedia Journalism. Last modified December 20, 2013. http://
mmj.vcu.edu/2013/12/20/methodology-gini-index-sidebar/.

"Maxima and minima." Wikipedia. Last modified October 9, 2016.
https://en.wikipedia.org/wiki/Maxima_and_minima.

Immelt, Jeffrey. "NYU Stern Graduate Convocation 2016: Jeffrey
Immelt." Filmed May 20, 2016. YouTube video, 18:27. Posted June
2, 2016. https://www.youtube.com/watch?v=hLMiuN8uSsk.

Erlanger, Steven. "'Brexit': Explaining Britain's Vote on European Union Membership." *New York Times*, October 27, 2016. http://www.nytimes.com/interactive/2016/world/europe/britain-european-union-brexit.html?_r=0.

Hardy, Quentin. "Cloud Computing Brings Sprawling Centers, but Few Jobs, to Small Towns." *New York Times*, August 26, 2016. http://www.nytimes.com/2016/08/27/technology/cloud-computing-brings-sprawling-centers-but-few-jobs-to-small-towns.html.

Acemoglu, Daron, and Pascual Restrepo. "The Race Between Man and Machine: Implications of Technology for Growth, Factor Shares and Employment." Unpublished manuscript, December 2015. https://pdfs.semanticscholar.org/4159/521bb401c139b440264049ce0af522033b5c.pdf?_ga=1.27764476.1700601381.1481243681.

Lemann, Nicholas. "The Network Man: Reid Hoffman's Big Idea." *The New Yorker*, October 12, 2015. http://www.newyorker.com/magazine/2015/10/12/the-network-man.

Romer, Paul. "Interview on Urbanization, Charter Cities and Growth Theory." Paul Romer (blog), April 29, 2015. https://paulromer.net/tag/charter-cities/.

Calmes, Jackie. "Who Hates Free Trade Treaties? Surprisingly, Not Voters." *New York Times*, September 21, 2016. http://www.nytimes.com/2016/09/22/us/politics/who-hates-trade-treaties-surprisingly-not-voters.html.

"Trans-Pacific Partnership." International Trade Administration, Department of Commerce, Washington, DC. Accessed December 9, 2016. http://www.trade.gov/fta/tpp/index.asp.

Index

Index

Index

Index

Index

Index

Index

Index

Index

About the Author

Satya Nadella is a husband, father, and the chief executive officer of Microsoft—the third in the company's forty-year history.

On his twenty-first birthday, Nadella immigrated from Hyderabad, India, to the United States to pursue a master's degree in computer science. After stops in America's Rust Belt and Silicon Valley, he joined Microsoft in 1992 where he would lead a variety of products and innovations across the company's consumer and enterprise businesses. Nadella is widely known as an inspiring, mission-oriented leader who pushes the bounds of technology while crafting creative and sometimes surprising deals with customers and partners globally.

Nadella's life is a journey of learning deep empathy for other people, which he brings into all he does personally and professionally. As much a humanist as an engineer and executive, Nadella defines his mission and that of the company he leads as empowering every person and every organization on the planet to achieve more. In addition to his role at Microsoft, Nadella serves on the board of directors for Fred Hutchinson Cancer Research Center and Starbucks. Satya and his wife, Anu, personally support Seattle Children's Hospital as well as other organizations in the Seattle area that serve the unique needs of people with disabilities. Nadella will donate all of his proceeds from *Hit Refresh* to Microsoft Philanthropies.